S0-EAH-601

STAR WARS

Alun Chalfont

STAR WARS

SUICIDE OR SURVIVAL?

LITTLE, BROWN AND COMPANY
Boston Toronto

FIRST AMERICAN EDITION

Library of Congress Cataloging-in-Publication Data

Chalfont, Arthur Gwynne Jones, Baron, 1919–
Star wars: suicide or survival?

Includes index.
1. Strategic Defense Initiative. 2. Ballistic missile defenses.
I. Title.
UG743.C47 1986 358′.1754 86–163
ISBN 0-316-13607-7

PRINTED IN THE UNITED STATES OF AMERICA

Contents

Diagrams

Figure 1: statement by Richard DeLauer, Under Secretary of Defense for Research and Engineering, *The FY 1985 Department of Defense Program for Research, Development, and Acquisition*, to the 98th Congress, 27 February 1984, pages 11–12. Figures 2, 4 and 8, reprinted by permission, copyright the *Washington Post*; figures 3, 5, 6 and 7 (drawn by Joe Lertola) copyright 1985 Time Inc., all rights reserved, reprinted by permission from *Time*.

Foreword

by George A. Keyworth Science Advisor to the President of The United States of America

In March 1983, Ronald Reagan changed the present course of twentieth-century Man. I make this rather bold assertion regardless of the eventual technical path of the Strategic Defense Initiative – or 'Star Wars' – which he announced that day. I make it because of a remark made to me by a senior European statesman: 'No matter the outcome of Star Wars, the nuclear era will never be the same again.'

In making his challenge to the scientific community, the President asked not just for an investigation into technology, but a re-investigation of the values and methods with which we defend our society. In reality, I predict that it will be the change in our present way of *thinking* – not the scientific gadgets – that will eventually be seen at the heart of this President's legacy.

History will record that the last half of twentieth-century Man saw weapons become available that – for the first time since the Old Testament – could destroy entire civilizations overnight. Worse, their devastation was totally out of proportion to any previous measure of military balance.

But in the beginning these weapons were few in number, their delivery was difficult and imprecise, and they had little military value for pre-emptive strike. Western policy thereby evolved about the rational avoidance of their use by assuring these weapons would survive any first strike – and total destruction of any attacker would immediately follow.

History will also record that twentieth-century Man gradually became more concerned about these weapons' survival than his own.

Starting in 1960, an almost frantic decade of arms control agreements culminated in the SALT I Treaty. By its own provisions this agreement was only a stop-gap measure establishing a framework. *Real* arms control was to take hold rapidly thereafter. Seven years were spent trying to produce SALT II. It failed – ostensibly because of intolerable Soviet actions such as Afghanistan. But there were deeper problems.

On the one hand there was an awakening that technology – and a massive Soviet strategic investment in ballistic missiles – had produced a situation vastly different from that of SALT I days. Technical advances in only the last fifteen years, especially in the ICBM, had led to real concerns for the prospect of an escalatory arms race, and an eroding stability between East and West. That the strategic balance had shifted was recognized. The only real discussion was over exactly when: 1978, 1979, or 1980? There was graver concern, however, that the underpinnings of stable deterrence had begun to shift as well.

They had begun to shift militarily as the 1970s closed and the ICBMs lost their survivability. The flexibility, payload, and accuracy of Soviet missiles had increased to the point that the West could find *no* survivable basing modes for its own missiles. The broader implications were that the even more fragile National Command Authority and bomber escape routes were also at risk. Worse, an explosion in data-processing coupled with this offensive power might produce the ability to localize – not necessarily locate – and effectively cripple the presently secure nuclear ballistic missile submarine force. The possible dramatic effects of espionage could of course never be predicted, but were always feared.

In 1982–3, a Presidential Commission on Strategic Forces assembled under General Brent Scowcroft, former National Security Advisor under President Ford, and foreshadowed these concerns. They pointed out that we in the West increasingly depend upon the threat of having to launch our forces under attack if they are to have any hope of survival. This increasing tendency to both shorter decision times and spasm warfare clearly indicated a tendency toward decreasing stability as well.

Simultaneously, the ethical underpinnings of pure offensive deterrence have come under fire. Scowcroft noted that:

> Deterrence is not, and cannot be, bluff. In order for deterrence to be effective we must not merely have weapons, we must be perceived to be able, and prepared, if necessary, to use them effectively against the key elements of Soviet power.

But this is the issue – bluff vs. planned use – that the Catholic Bishops develop as the central ethic against present policy. That they argue themselves into a paradox, i.e., nuclear weapons for deterrence are acceptable as long as you don't seriously think about using them, is at the heart of the problem. We find ourselves in a dilemma. To paraphrase Winston Churchill in an earlier age: We are riding to and fro upon a tiger we dare not dismount, and it's getting hungry.

There is an expanding popular confusion in the West as to just what comprises our justifiable self interest – evidenced by the broadly-based freeze movement. The unmistakable trend is that of decreasing popular support for future nuclear systems, both within nations and between nations. Western leaders cannot forever fend off such pressures. The trend is clear. Western strategic options are becoming alarmingly narrow. And at the same time, even Henry Kissinger points out that conventional arms control measures alone are insufficient, and we are at a justifiable 'impasse in thought' as just how to effectively catalyze really meaningful reduction levels.

Continuing upon the present course, with deterrence maintained solely by an ever-increasing offensive nuclear threat, and arms control at a stalemate, presents little hope and even fewer options. On the other hand, re-establishing the historical check-and-balance effects of defense vs. offense might well offer new avenues to restart the process to effective arms control. That is:

- to reduce the risk of war
- to reduce the arms of war
- to reduce the consequences of war.

Two years after President Regan assumed office, the Chairman of his Joint Chiefs of Staff – very concerned with the trend he saw for the West – recommended we adopt a new sense of 'strategic vision' to provide new options in the future. While obligatory modernization of US strategic offensive forces had already begun in 1981, these forces were seen to be most likely inadequate as we ended the next decade.

On 23 March 1983, the President proposed we explore the feasibility of a new strategy. Such a strategy would use greater reliance upon defense, rather than offense, to move toward a goal of someday making nuclear weapons *effectively* impotent and obsolete. He then proposed we initially explore the technologies to defend against the principal source of today's instability – the ballistic missile.

Such a strategy, even in its early stages, could restore stability: first by making pre-emption unfeasible, and ultimately by greatly reducing our reliance upon the weapons themselves.

In that same speech – often overlooked – the President noted that this would entail dramatically increasing our conventional leverage. If we are to change the course of an eroding stability we have to address the *complete* problem, not simply the veneer. It's my opinion, and that of many other scientists and professional military officers, that these goals can readily be met – and sustained – by the superior technical industrial bases of the free nations.

It is these larger implications of such a change from 'conventional wisdom' views of doctrine that initially presented a ripe target for the Western press and Soviet propaganda. But it has survived the criticism. Now two years later it receives increasing support from the public, the media, defense strategists, scientists, and even key political figures in the ranks of opposing political parties.

It is these larger implications of SDI that have redirected the initial querulous reactions to 'Star Wars', and motivated the Soviets to return to Geneva. Both sides are giving more serious discussion to the feasibility – and advisability – of the technologies and policies of SDI in a view toward a future that seeks to defend, not merely avenge, a free society. Moreover SDI appears to be the first enforceable catalyst to arms control in almost thirty years; and the only means by which I can foresee securing the implications of any truly drastic cuts in nuclear arsenals.

Lord Chalfont's contributions have been central to serious thinking on SDI since its inception. In this perceptive and incisive work, he analyzes the issue in a manner that reduces the tangled web of nuclear deterrence to the essential, and troubling, points that challenge today's leaders. He shares, in this short work, the logic that led him to conclude that the strategic implication of the SDI represents an important legacy from President Reagan to tomorrow's generation of leaders.

Preface

Any change in defence strategy must be managed with care and vision, and this is particularly so in the nuclear age. The Alliance is the best means to manage the course of change that the SDI embarks us on.

Dr George A. Keyworth

This brief study does not seek to make judgements about the science and technology of the Strategic Defence Initiative. Indeed, such entrenched positions have been occupied, not only by scientists but also by academic strategists, military professionals and specialist journalists, that it is not possible to intervene in this debate without some danger of being suspected of being disingenuous. Supporters of the Strategic Defence Initiative are regarded by its opponents as at best simple-minded romantics, and at worst hard-faced beneficiaries of the so-called military–industrial complex. On the other hand, the opponents of SDI are often branded as technological Luddites, or even 'useful idiots' giving aid and comfort to the Soviet Union.

At the heart of this divisive and often abrasive debate lies a new and controversial approach to the central political, military and ideological confrontation of the century: that between the Soviet Union and its allies on the one hand, and the United States and its allies on the other. It is therefore a matter of direct concern to everyone, not simply to those who are, or believe themselves to be, experts. It is important that every citizen who is at risk from the nuclear confrontation should understand what the Strategic Defence Initiative is. According to the President of the United States of America it is 'an effort which holds the promise of changing the course of human history'.

This, most people might agree, is no small matter; and the purpose of this book is to set out, in simple and largely non-specialized language, the major issues. It aims to describe the Strategic Defence Initiative; and to examine the arguments regarding its cost, technological feasibiiity and politico-stratetic implications. Most importantly, it attempts to place 'Star Wars' in the broader context of the strategic framework within which it was conceived – a concept which, if it can be realized, might provide for future world leaders a calculus of international relations which does not depend upon the threat of nuclear annihilation; which might, in fact, replace the suicidal strategy of 'Mutual Assured Destruction' with the prospect of mutually assured survival.

What is being contemplated entails high risk; arguably, however, no risk is higher than that which we face at present. It would be tragic if the prospects held out by the Strategic Defence Initiative were to be rejected because of a basic misunderstanding of what it is meant to achieve.

In preparing this analysis, I have been helped substantially by a number of friends and colleagues; they include General Sir Harry Tuzo, Dr G. A. Keyworth, Mr Herbert Meyer and Mr Gerald Frost. Lord Zuckerman and Dr Richard Garwin, two of the most distinguished and articulate critics of the concept of strategic defence, while not sharing my views on the Strategic Defence Initiative, have nevertheless offered wise and friendly advice. Valuable insights into the technology of strategic defence were provided by Dr Robert Buchanan and his colleagues at the BDM Corporation.

I am also grateful to the Editors of *The Times* and *Encounter* for allowing me to use material which originally appeared in their pages. Much of the material in Chapter 10 has been previously published in *Signal* of March 1984, the official journal of the Armed Forces Communications and Electronics Association (copyright 1984), and it is now reprinted by permission of AFCEA.

Finally, I would like to record my appreciation of the skill and patience of Angela Knowles, who has typed successive versions of this book.

Introduction

On 23 March 1983, in a televised speech, President Reagan announced that he was 'directing a comprehensive and intensive effort to define long-term research and development programmes to begin to achieve our ultimate goal of eliminating the threat posed by strategic nuclear missiles. . . . This would pave the way for arms control measures to eliminate the weapons themselves.' (See Appendix 1.) Subsequently, after preliminary investigations had suggested that research might prove that strategic defence of this kind was feasible, the President established a Strategic Defence Initiative Office (SDIO) under Lieutenant General James A. Abrahamson to implement the research programme.

This was, President Reagan insists, entirely his own idea:

> It kind of amuses me that everybody is so sure I must have heard about it, that I never thought of it myself. The truth is I did. . . . At one of my regular meetings with the (Joint) Chiefs of Staff, I brought up this subject about a defensive weapon . . . and I asked them, 'Is it possible in our modern technology of today that it would be worth while to see if we could not develop a weapon that could perhaps take out, as they left their silos, those nuclear missiles?' . . . And when they did not look aghast at the idea and instead said yes, they believed that such a thing offered a possibility and should be researched, I said, 'Go.'

In launching this programme, the Strategic Defence Initiative, the President addressed the American people with some vivid and cogent imagery:

> Let me share with you a vision of the future which offers hope. It is that we embark on a programme to counter the awesome Soviet

> missile threat with measures that are defensive. . . . What if free people could live secure in the knowledge that their security did not rest upon the threat of instant US retaliation to deter a Soviet attack; that we could intercept and destroy strategic ballistic missiles before they reached our soil or that of our allies? . . .
>
> Would it not be better to save lives than to avenge them? . . . My fellow Americans, tonight we are launching an effort which holds the promise of changing the course of human history. There will be risks, and results take time. But with your support, I believe we can do it.[1]

Even if one discounts the touch of drama that is understandable in a speech delivered by a political leader of President Reagan's background and temperament, it seems an attractive enough proposition. Since nuclear weapons were invented, there has been an almost incessant clamour from people demanding that something should be done about them. Suggestions have ranged from 'banning' them altogether – a concept which has about the same intellectual content as a proposal to ban mathematics or repeal the law of gravity – to 'no-first-use' declarations and the establishment of nuclear-free zones in places like Milton Keynes or the London Borough of Camden, presumably on the assumption that the designated area would automatically become invulnerable to blast and radiation.

It might have been expected, therefore, that when the leader of one of the two superpowers announced a programme ostensibly designed to make nuclear weapons obsolete, the reaction would have been generally favourable; but the path of the peacemaker is notoriously stony. The Soviet Union, of course, reacted with a characteristic sense of outrage. Within three days Mr Andropov, then President of the USSR, delivered his measured judgement. 'Let there be no mistake about it in Washington,' he said, 'it is time they stopped devising one option after another in the search for best ways of unleashing nuclear war in the hope of winning it. Engaging in this is not just irresponsible, it is insane.'[2]

In early 1984 a scientific broadside was fired from the

Soviet Union by a working group of the Committee of Soviet Scientists for Peace Against the Nuclear Threat. Its report concluded in effect that space-based defence systems would be too expensive, that they were technically unattainable, and that in any case they would be easily neutralized by countermeasures.[3] They went on to declare that

> . . . the assertions coming from the Reagan administration that the new anti-missile defence systems spell salvation from nuclear missiles for mankind are perhaps the greatest-ever deceptions of our time. The danger of the above-mentioned programmes is aggravated by the fact that they appeal to man's natural inborn desire to find a shield against the all-killing destructive power of nuclear weapons. . . .

The doubts thus cast upon the technological feasibility of strategic defence were apparently not shared by Mr Nikolai Basov of the Soviet Academy of Sciences, who announced in Moscow in January 1985 that the Soviet Union would have no technological difficulty in matching the American SDI programme, a claim which has since been endorsed by the Soviet Defence Minister, Marshal Sokolov.[4]

Ironically, however, many critics of the Strategic Defence Initiative in the West were at one with the peace-loving Soviet scientists. The popular press at once adopted the phrase 'Star Wars' to describe the initiative. It had an easy, familiar ring, as it was the title of a popular science-fiction film; it fitted easily into the headlines; and it treated President Reagan's concept with exactly the right kind of patronizing contempt – as a cinema-type fantasy, not to be taken seriously by responsible citizens. The Union of Concerned Scientists delivered an appeal to both superpowers to 'avoid an arms race in space'.[5] John Glenn, the Democratic Senator from Ohio and former astronaut, called for an agreement with the Soviet Union to ban space weapons. Such eminent American authorities as Mr McGeorge Bundy and Dr Harold Brown emerged as opponents of the SDI. In a lecture given to the California seminar on International Security and Foreign Policy in

November 1984, Dr Brown, who was US Secretary for Defense from 1977 to 1981, concluded that American policy ought to have two principal aims: to reduce the level of strategic offensive forces and promote the survivability of the force postures on both sides; and to ban all weapons in space while reaffirming the legitimacy of non-weapon space-based systems such as satellites for early warning, surveillance and communications. Carl Sagan, the prophet of 'nuclear winter', suggested that the most readily deployable ballistic missile defence would suffer from the disability, when it worked at all, of generating fires 'contributing to a climatic catastrophe'.

One of the most powerful and authoritative attacks on the Strategic Defence Initiative came in an article in *Scientific American* in October 1984. Signed by Hans Bethe, Richard Garwin, Kurt Gottfried and Henry Kendall, four American scientists of high reputation, it analyzed the technological implications of strategic defence and arrived at an unequivocal conclusion:

> In our view the questionable performance of the proposed defense, the ease with which it could be overwhelmed or circumvented, and its potential as an antisatellite system would cause grievous damage to the security of the US if the Strategic Defense Initiative were to be pursued. The path toward greater security lies in quite another direction. Although research on ballistic-missile defense should continue at the traditional level of expenditure and within the constraints of the ABM Treaty, every effort should be made to negotiate a bilateral ban on the testing and use of space weapons.

Meanwhile a broad spectrum of politicians, academic, strategic analysts and leader writers in the United Kingdom clearly regarded the Strategic Defence Initiative as a bad thing. Professor Lawrence Freedman conducted an intensive campaign against the idea, equalled in passion only by Mr Tam Dalyell's crusade against the Falklands war. One of his more extreme flights of rhetoric reflects Professor Freedman's vigorous, not to say violent, opposition to the SDI:

> President Reagan's speech of March 1983 may have launched a thousand research projects but it did not launch a strategic revolution. He was offering a false prospect of invulnerability, an illusion that he had some bold escape plan from the harsh realities of the nuclear age. This would have quickly been dismissed as the ramblings of a sentimental idealist had he not been President of the United States, and had he not backed up his vision with the promise of a technical solution that was soon found to be wanting.

Whatever might be said about President Reagan, Professor Freedman is evidently not a man intolerably plagued with self-doubt. Dr David Owen, the leader of the Social Democratic Party, addressed a long and closely argued philippic to Mrs Thatcher on the eve of her departure to the United States at the end of 1984, urging her to reject the SDI out of hand. She did not do so, but that is another matter. Edward Heath, the former British Prime Minister, described the programme as 'decoupling, destabilizing and a diversion of resources'.[6] Colonel Jonathan Alford of the International Institute for Strategic Studies was coolly sceptical of the whole concept while Lord Zuckerman, former Chief Scientific Adviser to the British Government, appeared to be resolutely opposed to it. These are not people whose views on matters of nuclear strategy can be lightly dismissed.

Possibly the most significant British political view, however, was expressed by the Secretary of State for Foreign Affairs, Sir Geoffrey Howe, in a speech delivered at the Royal United Services Institute for Defence Studies in London on 15 March 1985. Although the speech was drafted in the carefully coded language familiar to anyone who has experience of the British Foreign Office, its technique of praising with faint damns aroused considerable fury in Washington.

Sir Geoffrey began by rehearsing the four points of agreement which had been endorsed by President Reagan and the British Prime Minister, Mrs Thatcher, namely that:

> the US and Western aim is not to achieve superiority but to maintain balance, taking account of Soviet developments;

SDI-related deployment would, in view of treaty obligations, have to be a matter for negotiation;

the overall aim is to enhance, not undercut, deterrence;

and East–West negotiation should aim to achieve security with reduced levels of offensive systems on both sides.

He went on, however, to say that 'We must take care that political decisions are not pre-empted by the march of technology, still less by premature attempts to predict the route of that march' – a clear warning signal to the strategic defence scientists. He also questioned the cost and technological feasibility of the SDI and the wisdom of 'active defence', and the invulnerability of space-based defensive systems. Although the speech was presented as an objective analysis, by deploying all the familiar arguments of the nay-sayers it seriously undermined the Prime Minister's declaration that the United Kingdom was 'at one' with the US President on the SDI issue.

The Chinese adopted, not surprisingly, a 'plague on both your houses' attitude to the Strategic Defence Initiative. In an article in the *Beijing Review* in November 1984 Zhuang Qubing, a research fellow at the Beijing Institute of International Studies, reflected the official Chinese view. After suggesting that an American strategic defence system would leave the Western European allies vulnerable and that the equilibrium of military power would be destabilized, Mr Zhuang concluded:

> It's quite clear that the plan to establish a strategic nuclear weapon system in space is a dangerous one. Its main consequence will be even more tense US–Soviet relations, which will increase the danger of world war and threaten world peace. The development of space weapons would enhance the degree of accuracy of both nuclear and non-nuclear weapons, and would expand warfare in new directions. Some believe that should a war between the United States and the Soviet Union ever break out, it would start with an attack on the enemy's satellites.
>
> The rapid development of space technology is a great

> achievement of modern science. It can and should contribute to world peace and human progress. People all over the world demand that the United States and the Soviet Union stop the space arms race. In order to safeguard world peace, it is imperative to prohibit the development, testing, production and deployment of space weapons and to ultimately destroy all of them. Presently, the United States does not want to give in, and the Soviet Union is not sincere when it makes disarmament proposals. It is hoped that the two countries negotiate, come to really effective agreements, and cease such dangerous actions before an overall space weapon system is established. Stopping the militarization of space is in the best interests of the people of the United States, the Soviet Union and the rest of the world.

It was, however, among the so-called 'peace' movements of the West that the most sustained opposition to the SDI emerged. In parenthesis, it is of more than passing interest to note the propensity of protest movements for assuming a monopoly over convenient words and phrases. A 'peace' movement implies that anyone who disagrees with its aims and methods or impugns its motives is against peace; scientists who do not subscribe to the conclusions of Concerned Scientists are by definition unconcerned or unscientific. The principal message of the protest movements in the often confused debate over nuclear strategy has been the horror of nuclear war. In an attempt to play upon the understandable fear of the appalling devastation which would inevitably result from the use of nuclear weapons, the protest industry has persistently emphasized the aspects of death, mutilation, and devastation. It might reasonably have been expected, therefore, that they would extend, at the very least, a cautious welcome to a programme designed to eliminate nuclear war.

The knee-jerk reaction, however, is intellectually less demanding than the rational response. The Campaign for Nuclear Disarmament in the West at once attacked the SDI as an extension of the arms race and yet another manifestation of the trigger-happy, warmongering irresponsibility of the American administration and its President. The latent anti-Americanism which is never far below the surface in certain bands of the political spectrum in western Europe was brought out for yet another airing. The most bizarre aspect of the hostile reaction to the Strategic Defence Initiative was the closing of ranks around the concept of 'Mutual Assured Destruction'. This is the articulation of the form of mutual deterrence sometimes referred to as 'the balance of terror'.

Simply stated, it is the belief that, as long as each side in the nuclear confrontation has enough nuclear weapon systems to absorb a sudden attack by the other, and still retaliate with devastating effect, no one will start a nuclear war. It is a powerful argument; but it rests, obviously, upon the threat to destroy cities and kill millions of innocent citizens; for a retaliatory strike, by definition, could not be against the enemy's missiles, since he has already fired them. A corollary to this concept is that neither side should take any steps to protect its citizens against attack, since this would theoretically enable it to strike first without fear of retaliation. It is, therefore, a deterrent posture based upon the implied readiness to commit collective suicide rather than surrender.

The Strategic Defence Initiative was the catalyst for a strange display of affection towards this somewhat macabre concept. Monsignor Bruce Kent, then General Secretary of the Campaign for Nuclear Disarmament, said in an interview with the London correspondent of *Pravda*, 'Any steps aimed at averting a nuclear strike inevitably weaken the deterrent element in the existing nuclear balance'[7] – a classic statement of the MAD apologia. Faced with a strategic theory which offered survival as an alternative to suicide, there was a strange disinclination to abandon the familiar certainties of annihilation. It is therefore important that there should be a greater degree of clarity about the issues of nuclear strategy, and especially of strategic defence. What is the present state of the balance of forces? Is Mutual Assured Destruction a valid concept? What does the Strategic Defence Initiative really mean in terms of effective defence, its impact on arms control and its potential for making nuclear weapons obsolete? Before any useful assessment of SDI can be made, a calm reappraisal of the facts might be in order.

To put it mildly, 'Star Wars' has had a bad press. In considering this phenomenon it is important to establish a frame of reference by means of a brief résumé of the Strategic Defence Initiative, since it is clear that some of the arguments advanced against the idea can be explained only by the existence of a serious misunderstanding of what it actually is. Only then is it possible to examine some of the more substantive objections to SDI and to test their validity, and perhaps to advance the proposition that SDI is a concept which deserves a much more sympathetic reaction than has so far been evident – especially in Western Europe.

It is, however, not possible to approach the specific issue of ballistic missile defence without placing it first in the wider political and strategic context.

1 The Nuclear Dilemma

One of the fallacies at the heart of the universal concern about nuclear weapons is the belief, carefully fostered by the 'peace movements', that we are on the brink of a terrible holocaust, which will lay waste to the world, annihilate its inhabitants in millions and bequeath to future generations a legacy of mental and physical disease. That is certainly what would happen if there were a major nuclear war; but there is probably less danger of war now than there was on many occasions before the nuclear weapon was invented. The evidence of history is that arms races alone do not cause wars. The causes of war are subtle and intricate; they derive from political, economic and even psychological forces which eventually lead a power or group of powers to calculate that by using its military strength it can achieve gains that will outweigh the costs of war.

As far as the major armed confrontation in the world today is concerned – that between the Soviet Union and its allies on the one hand and the Western Alliance on the other – the advent of nuclear weapons has profoundly and irrevocably altered the calculus. There is no political prize which either side could conceivably regard as being worth the risk of a nuclear exchange. War has always been a form of lunacy; what has changed is that, whereas it has often in the past been undertaken by politicians displaying every appearance of sanity, today, in the world of nuclear weapons, it would be embarked upon only by certifiable lunatics. This is, of course, demonstrably not true outside the major East–West confrontation where nuclear weapons are not involved; but now that the great powers have the capacity to inflict intolerable damage and suffering upon each other, *and as long as each side knows that the other possesses that power*, the possibility of war between them is in fact remote.

This is not to suggest that such a situation is entirely desirable or even acceptable. Madmen have risen to positions of absolute power before, and they may do so again. The significance of the current uneasy confrontation is that, if the superpowers were plunged into a war, the consequences for the rest of the world would be almost unimaginably dreadful – partly because they have amassed a stockpile of weapons with such potential for destruction that few people anywhere would escape the effects entirely. The problem we face, therefore, is how to reduce those stockpiles to a level at which the capacity to deter a potential enemy from going to war remains, without posing a threat of total global disaster if deterrence should ever fail. That is the heart of the nuclear dilemma.

The present array of nuclear armaments, whatever criteria are used to calculate it, is formidable; and it is difficult to strike a balance which has any real meaning, partly because the inventory is constantly changing, and partly because any assessment of numerical superiority depends on what is being counted – the number of missiles or 'delivery vehicles'; the number of warheads; the 'throw weight' (the weight of the warheads, guidance systems and penetration aids which can be delivered to a given target); or the 'yield' (the destructive power expressed in terms of the equivalent tonnage of conventional high explosive). For example, the balance of intercontinental or strategic weapons is substantially in favour of the Soviet Union when measured in delivery systems (2,387 Russian to 1,572 American); but in favour of the Americans in the number of warheads (approximately 7,000 to 5,000). The Soviet Union has a substantial superiority both in throw weight and in total yield. Perhaps, however, the most significant statistical fact is that between them the Soviet Union and the United States are capable of delivering, using their strategic systems alone, the nuclear equivalent of between three and four thousand million tons of high explosive.

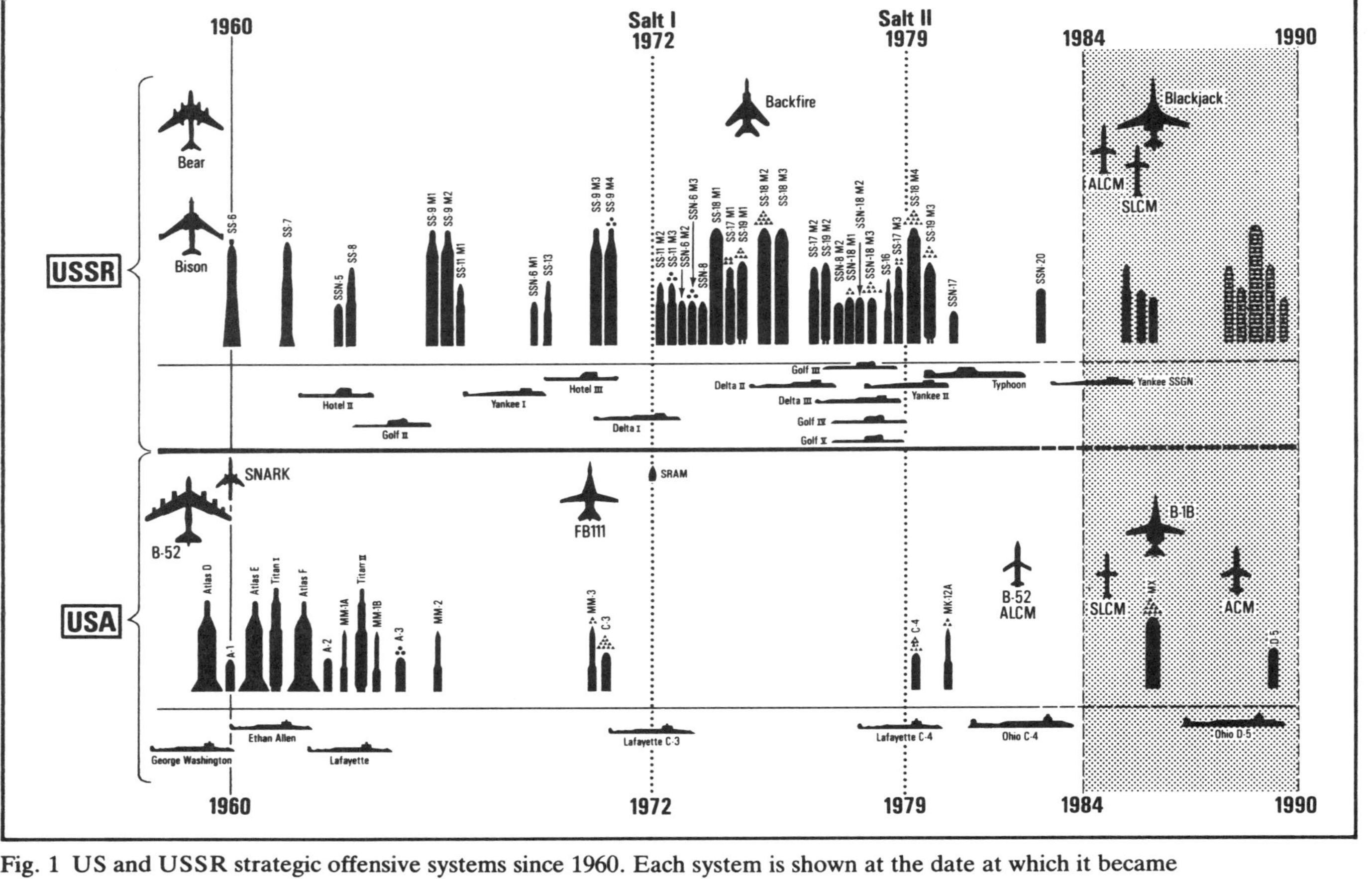

Fig. 1 US and USSR strategic offensive systems since 1960. Each system is shown at the date at which it became operational. Dots above missiles indicate the number of warheads.

Although it is possible to argue that the Soviet Union is beginning to enjoy an overall measure of advantage in strategic nuclear weapons, the situation is complicated by other factors. Two of America's allies, France and Britain, have their own strategic weapons aimed at targets in the Soviet Union; no other member of the Warsaw Pact has its own nuclear weapons. Furthermore, there is now a great array of intermediate-range and 'tactical' or 'battlefield' nuclear weapons. The Warsaw Pact has a massive superiority in intermediate-range or 'theatre' nuclear weapons, in tactical weapons and in aircraft capable of delivering nuclear weapons; while NATO has a considerable edge in atomic artillery. Perhaps the most significant development, however, lies in the future.

During the period of the Carter presidency in the late 1970s, American development of nuclear weapons systems was substantially reduced, whereas in the Soviet Union it continued unabated. Given the time needed to bring these systems from the drawing board to operational readiness (the 'lead time'), this is likely to give rise to a situation in the late 1980s in which the Soviet Union might enjoy a measure of superiority. This period, known in the jargon as the 'window of vulnerability' is one in which, according to some American strategic analysts, the Soviet Union would theoretically be able, in a surprise nuclear attack, to destroy a substantial proportion of American land-based strategic missiles, using only a small number of its own, thereby leaving the United States facing massively superior Russian nuclear forces in any subsequent exchange. However theoretical and unreal this calculation might seem, it understandably exerts a powerful psychological effect on American planners.

The position has, therefore, changed dramatically over the years since World War II. At first the United States had a complete monopoly of atomic weapons, a position which enabled it to threaten the Soviet Union with virtual annihilation in response to a Russian military attack of any kind. This was the era of 'massive retaliation', a doctrine which ceased to be valid as soon as the Soviet Union, in the

1950s, acquired its own atomic weapons and, more significantly, the capacity to absorb an American nuclear attack, and still retaliate with its own nuclear weapons – the 'sure second-strike capability'. Subsequently the Russian nuclear armoury has grown consistently, until in the late 1970s the Soviet Union could match the United States in the total number of nuclear delivery systems. If the 'window of vulnerability' theory has any validity, the Russians might have, by the end of the 1980s, not just a sure second-strike capability, but a theoretical ability to launch a surprise nuclear attack on the United States – the 'pre-emptive' or 'first-strike' capability.

Meanwhile, Russian military strength has been growing in other respects as well. The Soviet Navy has developed, in thirty years under the formidable Admiral Gorshkov, from a relatively innocuous force consisting principally of coastal defence forces into the second most powerful ocean-going navy in the world. Furthermore, this great maritime force has, together with the Soviet Air Force, provided the Soviet Union with the capacity to move military forces and equipment rapidly and effectively to almost any point on the globe – the 'distant application of military force'. Finally, the balance of land-based military power in Europe has, over the past twenty years, moved steadily in favour of the Warsaw Pact, while the West has largely lost the technological edge which allowed NATO to calculate that quality was more important than quantity.

The growing military strength of the Soviet Union, together with a relative decline in the Western position, has inevitably been reflected in the changing spheres of world influence. In the last twenty-five years a Russian presence of one kind or another has been firmly established in South-East Asia (Vietnam and Cambodia); in the Middle East (South Yemen); in North Africa (Ethiopia, Libya and Algeria); in southern Africa (Angola and Mozambique); more recently in Central and South America (Guatemala, Nicaragua and El Salvador); and of course in the most dramatic way in Afghanistan, which has been under Soviet occupation since the end of 1979. As the United States Defense Secretary, Caspar Weinberger, has said,

> There is nothing hypothetical about the Soviet military machine. Its expansion, modernization and contribution to projection of power beyond Soviet boundaries are obvious. A clear understanding of Soviet Armed Forces, their doctrine, their capabilities, their strengths and their weaknesses is essential to the shaping and maintenance of effective US and Allied armed forces.

As the will and the capacity of the Soviet Union to project its military power beyond its own borders has manifestly increased, so that of the United States and the West has declined. Since the end of the Vietnam war, American power has been progressively withdrawn from the Far East into the European Atlantic area, and the violent popular reaction against the involvement of the United States in Vietnam has led to a reluctance on the part of successive American administrations to become entangled in military situations remote from their own immediate spheres of influence. As a result of this, the United States has taken no direct action to counter Russian penetration of southern Africa. When the Shah of Iran was deposed and the staff of the United States Embassy in Tehran were taken prisoner by Khomeini's mob, the only American reaction was a humiliatingly abortive rescue attempt; and when the Red Army moved into Afghanistan there was no effective Western response. At the same time Britain has, under inexorable economic and social pressures, abandoned its own global role. Since the disastrous Suez operation of 1956, British forces have been progressively withdrawn from most of their overseas bases; and the only sign of a residual determination to resist aggression came in 1982, when Argentine forces invaded the Falkland Islands and were summarily ejected by a British task force.

On the other hand, the Soviet Union, apart from increasing its presence and power all over the world, has reacted vigorously to any threats to its own area of influence. In East

Germany in 1953, Hungary in 1956, Czechoslovakia in 1968 and Poland in 1980, Russian armed force, or the threat of it, was used ruthlessly to put an end to any sign of unravelling in the Soviet Empire. The post-war story has been one of inexorable Russian expansion. It is true that there have been setbacks. In 1962 President Kennedy frustrated, by a display of naval power and iron nerve, an attempt to emplace Soviet missiles in Cuba (and, by doing so, set in train the development of the modern Soviet Navy). In 1972 President Sadat of Egypt expelled thousands of Soviet military advisers, causing the Russians to comment that 'the leadership of Egypt had abandoned the tradition which the Egyptians had pursued for many years'. In 1973, when American intelligence sources reported that Russian intervention in the Egypt-Israel war was imminent, United States forces, including the strategic nuclear strike force, were placed on 'red alert', and the threat evaporated.

The background to the current nuclear dilemma has therefore been one of persistent confrontation between the great superpower groupings – a confrontation moving uneasily between episodes of crisis and periods of so-called 'détente', in which the struggle for predominance shifted from the military dimension to that of diplomacy and economic strategy. The competition in nuclear weapon systems has in the meantime been virtually unceasing. As the stockpiles have grown, and the accuracy, destructive power and technical sophistication of the systems has increased, concepts of nuclear strategy have undergone a significant transformation. In the 1960s and early 1970s there was a clear, if tacit, understanding on both sides of the confrontation that nuclear weapons were, in the last analysis, unusable. Their principal function was as a mutually cancelling deterrent, maintained solely to insure each side against attack by the other. This situation was known variously as the 'balance of terror', the 'nuclear stalemate' or, by strategic analysts, as 'Mutual Assured Destruction' (usually abbreviated, much to the delight of the anti-nuclear lobby, to the acronym MAD).

With the emergence of new weapons of great accuracy and penetrative power, strategic theorists have more recently begun to discuss the possibility of 'disarming strikes' in which nuclear missiles would be used not to threaten the indiscriminate destruction of the enemy's cities, but to destroy his capacity to retaliate. Thus it is sometimes suggested that a nuclear war might now be 'winnable'. Variations on this bizarre theme include speculation about the possibility of a 'limited' nuclear war, in which an exchange of nuclear weapons between adversaries might be restricted as to their number or type, or the geographical area in which they were used. One of the results of this strange excursion into fantasy has been to exacerbate the natural fear of nuclear war and to provide a powerful impetus to anti-war and anti-nuclear protest movements in the West. Meanwhile, however, the built-in momentum of nuclear weapons development continues. Since late 1981, the Soviet Union has begun test flights of two new land-based intercontinental ballistic missiles, a new generation of strategic manned bombers and a new series of cruise missiles. President Reagan has authorized the deployment of a new intercontinental system, MX, and the United States has, with the consent of its allies, deployed cruise missiles and a new generation of intermediate-range ballistic missiles in Western Europe.

This constant accretion of nuclear weapons – sometimes referred to as 'vertical proliferation' – is now in danger of being matched by the development of 'horizontal proliferation' – the spread of nuclear weapons outside the five major powers which now possess them, namely the United States, the Soviet Union, the People's Republic of China, Britain and France. A number of countries possess the scientific and technological skills, together with the essential industrial base, to become nuclear-weapon powers. India has already tested a nuclear device and is a nuclear-weapon power in all but name. Other countries such as Israel and South Africa are believed to have developed a nuclear-weapon capacity in every respect short of an actual explosion (and some reports suggest that South Africa may already have conducted a secret test). Iraq and Pakistan have highly developed nuclear programmes, and others such as Australia, Japan and Argentina already have a clear potential nuclear capability. It is, in fact, within the capacity of any highly industrialized country with access to fissile material to make and successfully test a nuclear weapon within a relatively short time.

In an attempt to eliminate the obvious dangers of horizontal proliferation, a treaty was concluded in 1967 and subsequently signed by and ratified by more than a hundred states. Under the terms of the Nuclear Non-Proliferation Treaty, nuclear-weapon states agree not to transfer, and non-nuclear weapon states not to receive, nuclear weapons or the technology for their manufacture. Under Article 6 of the Treaty, however, the signatories (including the Soviet Union, the United States and Britain) undertook 'to pursue negotiations in good faith on effective measures relating to cessation of the nuclear arms race at an early date'.

The Non-Proliferation Treaty is only one example of the numerous attempts which have been made to control the power of the nuclear weapon since it made its first appearances at Hiroshima and Nagasaki. As early as 1946 the United States government put forward the Baruch Plan for the establishment of an international authority to control all atomic fuel and facilities and to supervise the destruction of all nuclear weapons. The Soviet Union rejected the Baruch Plan and put forward one of its own which was rejected by the United States on the grounds that it made inadequate provision for international inspection – an issue which has bedevilled arms control negotiations ever since. In 1959 the Antarctic Treaty prohibited all military and nuclear activities in the area; in 1963 the Partial Nuclear Test Ban Treaty banned the testing of nuclear weapons in space, in the atmosphere and underwater (but not underground); in 1967 a treaty was signed prohibiting the stationing of missiles in outer space; and in 1971 the positioning of nuclear devices on the seabed was forbidden.

In 1971 the first of the Strategic Arms Limitation agreements (SALT I) sought, without great success, to place limits on both the offensive and defensive nuclear missile systems of the United States and the Soviet Union. In 1974 the two superpowers agreed to the cessation of underground nuclear tests above a yield of 150 kilotons (the equivalent of 150,000 tons of high explosive); and in 1979 a second strategic arms limitation treaty (SALT II) made another attempt to limit American and Russian delivery systems. It has, however, never been ratified.

Whatever the value of these various agreements might be (and it is certainly greater than radical disarmament movements suggest), they suffer from two major weaknesses. The first is that they have never been universal in their application. Two of the five nuclear powers (France and China) have taken no part in the negotiations and are not bound by the treaties. The second is that no provision has been made for actual arms reduction: for the reduction of existing levels of nuclear weapons. This has now become a matter of great urgency. Although the 'numbers game' – the attempt to judge the nuclear balance on a numerical basis – is academic, and, in the context of the massive destructive power on both sides, largely irrelevant, a continuing competition in nuclear arms has a number of inherent dangers. In the first place it distorts national economies by diverting valuable technological and human resources; secondly it perpetuates a situation in which mechanical accident or gross political misjudgement could have catastrophic consequences; and thirdly, and most important of all, an upward spiral of nuclear weapons development poses the threat that, at some stage, one side or the other may perceive that it has 'superiority', or that the other side might be about to achieve it – the 'window of vulnerability' theory. At such a stage the danger of a surprise or 'pre-emptive' attack is substantially increased.

A new and equally destabilizing factor has been introduced into the equation by the recent growth in the West of radical protest movements demanding one-sided disarmament and the dismantling of the Western Alliance. The inescapable consequence of what they are demanding would be to make nuclear war more, rather than less, likely. An uncontrolled, unbalanced *downward* spiral of disarmament would create the same dangerous instabilities as an uncontrolled upward spiral. Furthermore, unilateral disarmament by the West would clearly remove any incentive which at present exists for the Soviet Union to engage in binding international agreements.

There are signs that the great powers are at last beginning to approach with a properly serious sense of purpose the problem of resolving the nuclear dilemma. Since 1981 the Soviet Union and the United States have been discussing the problem of intermediate-range nuclear forces (INF), in an attempt to reduce the number of missiles designed for 'theatre' use – that is to say, designed to destroy military installations and other targets in Europe. The major threat to the West in this context is the Russian SS-20 – a powerful, accurate and mobile missile system of which over 200 are now aimed at Western Europe. In an attempt to counter this threat NATO has begun to station American cruise missiles, as well as a new version of the Pershing ballistic missile, in certain countries of the Alliance (a development which has provoked increased activity among the protest movements).

In the INF talks the United States has now proposed the so-called 'zero option', under which the NATO deployments would be abandoned in exchange for the Russian undertaking to destroy all its SS-20s. The Soviet Union has rejected the idea, demanding instead that British and French nuclear weapons should be included in any INF agreement. Meanwhile, the old Strategic Arms Limitation Talks have, significantly, been renamed START (Strategic Arms *Reduction* Talks). When they opened in Geneva in June 1982, the United States immediately proposed reductions of nuclear warheads by one-third and of ballistic missiles by a half, with further reductions to follow. So far the Soviet Union has been unable to agree.

The process of bringing the nuclear weapon under control is likely to be a long and difficult one, requiring the constant exercise of political wisdom and intellectual subtlety. The facile slogan 'Ban the Bomb' serves only to exacerbate irrational fears and stimulate unrealistic hopes. The knowledge of how to construct, test and use nuclear weapons exists; and there can never again be a certainty that, even if all existing nuclear weapons were destroyed, someone in the future, faced with the prospect of defeat in a conventional war, would not again construct and use a nuclear weapon. Furthermore, even if it were possible for the world to engage in some collective act of self-induced amnesia, and banish nuclear technology from the human consciousness, the world would not necessarily be a safer place. We would all then be at the mercy of any aggressor with powerful 'conventional' armed forces, undeterred by the prospect of nuclear devastation. Those who declaim passionately on the horrors of nuclear war tend to forget that 'conventional' war too is cruel, destructive and barbaric.

The nuclear dilemma contains an inherent paradox. The nuclear weapon has injected a new dimension into the concept of war as a means of settling international disputes. Nuclear war would bring with it such appalling devastation that no nation state could conceivably hope to achieve any proportionate political aim by embarking on it. As long as nuclear weapons exist, furthermore, political leaders are less likely to engage in a war *of any kind* for fear that it might 'escalate' into a nuclear one. The problem is to devise a system of deterrence at substantially lower levels so that, in the dreadful event that deterrence should fail, the result would not be universal annihilation. This state of affairs is more likely to be achieved at negotiating tables in Geneva than in the streets of London or Berlin. The background of the nuclear dilemma brings out the sombre truth that in the complicated world of international relations there are no 'quick fixes' or easy solutions. This does not, however, absolve international statesmen from the need to search for long-term alternatives to the present dangerous confrontation.

2

The Political Confrontation

If the free world is ever enslaved, it will be, to a great extent, because it has lost the will to remain free. Indeed there are even now those who question the concept of freedom, pointing out that it is differently interpreted in the Western and Communist worlds. From this irredeemably confused intellectual approach emerges the fallacy of moral equivalence – a suggestion that there is some kind of symmetry between the aims of policy of the two great superpowers, and that they are equally responsible for the East–West confrontation and the threat of nuclear war which arises from it. E.P. Thompson, the Marxist historian and leading intellectual guru of unilateralism, has described the confrontation as 'a degenerative state' driven by 'an interoperative and reciprocal logic which threatens all impartiality'.

It is possible to argue that this attempt to identify a position of neutrality between the Soviet Union and the West is as perverse as to suggest that there is a position of neutrality between the arsonist and the fire brigade. Some things, it might reasonably be supposed, are evil and cruel and ugly, and no amount of turgid philosophical jargon will make them good or kind or beautiful. The East – that is to say in this context the Soviet Union, its clients and allies – endorses and preserves a political system in which individual liberty is accorded lower priority than the interests of the state. The basic political freedoms – of speech, assembly, religion and expression – are either denied or severely restricted. The basic civil liberties – equality of opportunity and equality before the law – are contemptuously ignored. Political dissent is ruthlessly suppressed. Indeed, to disagree openly with those in authority is held to be evidence of mental derangement, to be dealt with by 'psychiatric treatment'.

The nations which comprise the Soviet Empire live not only in a state of economic and political serfdom, but in what for some people is even worse, a condition of cultural barbarism which has been vividly described as 'historical amnesia'. Men and women are systematically cut off from their literary and artistic roots, from any cultural or spiritual stimulus that might provoke them into resisting the relentless oppression of totalitarian Communism. At the ideological centre of Soviet imperialism there still lies the dogma of Marxism–Leninism, from which is derived the belief that the Communist system will eventually prevail, and that anything, however evil, cruel and ugly which accelerates that process is by definition morally defensible.

The West, a much looser grouping of nation states comprising the United States and its clients and allies, generally speaking bases its political arrangements on the primacy of individual liberty and the conviction that the state exists to serve the citizen, not the citizen to serve the state. It would be idle to pretend that there are not deficiencies and injustices in the Western system. There is poverty, prejudice and inequality; but there are no *gulags*, no psychiatric wards for political dissidents, no pervasive secret police. There is a measure of cultural debasement, but it is self-inflicted, not systematically imposed from above. In preserving the delicate balance between freedom and order which is at the heart of political and social organization, the West leans, sometimes perhaps a little too heavily, in the direction of freedom. It is, however, not possible to enjoy freedom without a measure of order, while it *is* possible to impose order without concern for freedom. This is the basic difference between the political systems of the Soviet Empire and the West.

It needs to be stated unequivocally that these two systems are, and always will be, irreconcilable. That is not to say that they might not, in certain circumstances, exist side by side; that it might not be possible one day, as Winston Churchill said thirty-five years ago, to 'bridge the gulf between the two worlds so that each can live their life, if not in friendship, at least without the hatreds of war'. This seems, however a remote possibility in the context of Lenin's bleak prophecy that, one day, a dirge would be sung either over capitalism or over the Soviet Union. For the present and the foreseeable future it would be wiser to arrange our affairs to take account of the world as it is, rather than the expectation of the world as it might be some day.

In the real world, the Soviet Union has maintained massive military forces continuously since the end of World War II; there was no dismantling of the apparatus of war as there was in the West. More recently the Russian military machine has developed into a massive and potentially overwhelming instrument of foreign policy; and that foreign policy has been directed clearly and unmistakably at the progressive expansion of the Soviet Empire. As a result of this, the free world has concluded that its survival is at risk. It has nevertheless been fashionable in some quarters to dismiss the 'Soviet threat' as an invention of the febrile imagination of cold warriors, encouraged by the 'military-industrial complex'.

One of the most significant aspects of the contemporary political scene in Western Europe is a pervasive and often virulent anti-Americanism. The banners and placards of the columns which march earnestly through the streets of our cities under the title of peace movements proclaim a variety of messages . . . anarchist, Communist and pacifist. Many of them demand the removal of American bases from Europe, evidently under the impression that our cities are under threat from Pershings and not from SS-20s. This is depressing, but not surprising. For some time it has been possible to discern, in the pronouncements of the more fashionable soothsayers of press, radio and television, a magisterial tendency to adopt a position of spurious objectivity as between the free world and the totalitarian ideologies which threaten it. The confused rhetoric deployed in the press on the subjects of the cold war, the nuclear arms race, détente and the North–South dialogue has succeeded in anaesthetizing and indoctrinating a whole generation so that it has become incapable of distinguishing between reality and fantasy. Writing of the younger generation Mr Gerald Kaufman, a Labour politican not by any means of the extreme left, says that

> . . . defence policy cannot be justified to them as essential for deterring the Russians who . . . have never presented themselves to these teenagers as a direct threat to Britain's survival. To these young people, a greater justification for an independent British nuclear capability than a theoretical Soviet threat will have to be proved. . . .

One is driven to ask, how then can defence policy be justified to them, and what greater justification than the 'theoretical' Soviet threat could possibly be advanced for Britain's nuclear deterrent? The logic of this demoralizing argument is that young people are not only horrified by the possible effects of a nuclear war (a reaction which they share with their elders); they have also apparently concluded that this postulates unilateral disarmament for Britain – not only nuclear, but conventional as well. If that is so, it is largely because publicists of Mr Kaufman's persuasion have done nothing to expose the fallacy.

The next step is predictable, indeed inevitable. Once the 'plague on both your houses' heresy is established in the public mind, it no longer seems perverse to suggest that the threat to world peace comes not from an aggressive and expansionist Soviet Union, or even from an uncontrollable military competition between two demented superpowers, but from the United States of America and its trigger-happy President. Indeed a television critic writing in a national newspaper, evidently writing with the gears of his mind disengaged, recently declared that a programme presented by one of our more progressive television journalists had demonstrated that '. . . the real threat to peace comes not from the Kremlin, but from the man in the cowboy suit.'

Leaving aside the psychopathology of the endless cheap jibes at President Reagan's background, it is interesting to trace the aetiology of the new crusade against the Americans. Partly, of course, it springs from the instinctive hostility of the weak towards the strong, the poor towards the rich, the vacillating towards the resolute and decisive. It is interesting to observe that during the presidency of Mr Carter, while the global reputation of the United States was under attack from many quarters, anti-Americanism took a holiday. It is true, of course, that President Reagan's occasional tendency to make statements which have not been passed through a series of public-relations filters provides his enemies and those of his country with excellent ammunition.

When he says that it is possible to envisage a limited nuclear war in Europe, the pacifist–neutralist industry at once sets up a howl of execration. Yet he was only enunciating, possibly with imprudent clarity, the assumption upon which the defensive strategy of NATO has been based for twenty years.

If the doctrine of flexible response or graduated deterrence means anything at all, it postulates a possible sequence of events in which the Soviet Union launches a conventional military attack on Western Europe; inferior NATO conventional forces are incapable of containing it; 'battle-field' or 'tactical' nuclear weapons are used; whereupon the Soviet Union, convinced of the determination of the West to resist, abandons its aggression. In other words there has been a nuclear exchange limited to the European theatre. This may not be a very convincing basis upon which to construct a defensive strategy. Indeed, it is very unlikely that a war ever could be limited in this way. It seems, however, perverse to vilify the President of the United States because he is indiscreet enough to identify the underlying assumption of NATO's collectively agreed defensive strategy.

It is interesting, too, that in the uproar which greeted the American President's unguarded comment on the possibility of limited nuclear war in Europe, one important fact seemed to go unremarked: namely, that a war of *any* kind in Europe will occur only as a result of Russian aggression; and as Russian strategists have never left any doubt that they regard nuclear weapons as legitimate instruments of war, concern about NATO's plans for limiting their use seems curiously misplaced.

We have therefore a most alarming state of affairs. Not only has the Western Alliance blundered into a persistent state of crisis; it has also persuaded itself that most, if not all, of the fault lies with the Americans. It is therefore not only acceptable, but in some circles almost mandatory, to characterize Americans as naïve and dangerous militarists, propping up unsavoury dictatorships, provoking and perpetuating a mindless arms race, and attempting to bully their more civilized and sophisticated European allies into accepting a Manichean view of the world in which the only enemy is Communism and the only way to defeat it is 'the Bomb'. The Strategic Defence Initiative has produced another twist in this spiral of confusion and mistrust.

It would be prudent for those in Western Europe, where the record of wisdom in international affairs is by no means a matter for uncritical admiration, to pause before criticizing the Americans and their President. In Britain, especially, this particular brand of patronizing hostility is notably unbecoming. When we were recently engaged in a confrontation with Argentina, which culminated in a vicious little war, the behaviour of the United States was crucial. They began by making a series of careful attempts at mediation, based upon the justifiable calculation that America was the only nation with both the power and the credibility to bring about a peaceful solution; and when mediation failed, they came down unequivocally on the side of their NATO ally, providing intelligence, logistic reinforcement and moral support on a scale which was certainly influential, if not decisive, in the eventual outcome of the war.

It is also important to recognize that this took place against the background of a shift of emphasis in American strategic perceptions which is of primordial importance for the future of the Western Alliance. For over thirty years, United States foreign policy has rested securely on the assumption that, to use a somewhat melodramatic conceit, the River Elbe is one of the frontiers of the United States. In other words, there has been an almost unchallenged consensus that one of the critical areas of American security lies in Europe and the Atlantic. It is upon this foundation that the North Atlantic Treaty Organization was built; it is for this reason that the United States keeps 300,000 troops in Germany, partly as an element of conventional military defence, and partly as a 'hostage force' to demonstrate the validity of the American nuclear guarantee.

In recent years much has happened to cast doubts upon this central set of assumptions. Many percipient observers decline to believe in the credibility of 'extended deterrence', and question the validity of the theory of the American nuclear umbrella. To many Americans, on the contrary, it poses the very real danger that some day an American President might have to respond to a Russian attack on West Germany by 'pressing the button' and initiating a nuclear exchange in which American cities would almost certainly be included among the targets. Furthermore, the Soviet threat has long since ceased to be a matter of the red hordes pouring through France and Germany 'down to the Channel ports'. Vietnam, Afghanistan, Yemen and Angola have demonstrated the global reach of Russian foreign policy, powerfully supported by maritime forces which have developed in thirty years to provide not only a formidable navy presence across the sea lanes of the world, but also a capacity for the distant application of military force which is probably unmatched by any other world power.

The latest development in this context has been the emergence of a pattern of Communist penetration in Latin America. For many Americans, events in Nicaragua and El Salvador are seen as the early warnings of a political transformation which might eventually pose a direct threat to the security of the southern United States. One American reaction to this new perception is a demand, growing in insistence, for the withdrawal of American forces from Europe – a 'bring the boys home' movement which, however precarious may be its basis in military logic, has a powerful emotive appeal. At a more sophisticated level, the concept of 'global unilateralism' is gaining in strength. In essence, this postulates an American foreign policy liberated from entangling and institutionalized alliances, free to seek its friends and allies pragmatically and to construct security policies with more relevance to the pervasive and global nature of the threat. It is a modified form of isolationism which would not necessarily exclude defence arrangements with *some* Western European countries, but which would almost certainly signal the end of the North Atlantic Treaty Organization in its present form.

It might at this stage be salutary to point out that if that happens, one of the cardinal aims of Soviet foreign policy will have been achieved. It is a sterile exercise to engage in a debate about whether the Russians have a 'master strategic plan' or whether their policies are governed by a flexible opportunism. The dangers for the free world are the same in either case. What is clear beyond any reasonable doubt is that Russian long-term aims contain a number of identifiable elements. One of these is to bring about the disintegration of NATO and the separation of Western Europe from the United States as a necessary prelude to the Finlandization of Western Europe. (The Finns understandably resent the term, but perhaps they will permit its use in this context to identify in shorthand a political situation which is fully understood by the rest of the world.)

The Soviet Union therefore has good reason to be pleased with what is happening inside NATO. The United States, with the approval of most of its allies, and indeed at the express request of some, makes plans for the deployment of enhanced radiation warheads in Europe; there is, at once, an uproar against the 'neutron bomb' – described by the mentally enfeebled as a capitalist weapon, designed to kill people but preserve property. In an attempt (however misguided) to redress the geopolitical impact of the Russian SS-20 missile, it is proposed to station cruise missiles in certain Western European countries. Immediately the peace industry begins its predictable uproar. Most recently the

Strategic Defence Initiative and the reaction to it in Europe have added fuel to the great bonfire of suspicion, misunderstanding and open hostility which is being carefully kindled between the United States and its European allies.

The rise of neutralism and anti-Americanism in Western Europe has an especially piquant flavour when considered in the context of the confrontation between the world's two principal political, ideological and economic groupings. The Soviet Union is a police state controlled by a totalitarian dictatorship; denial of human rights is institutionalized and any sign of dissent is brutally suppressed. Its central assumption is that the human individual exists to serve the state, and not the state to serve the individual. Freedom has been to generations of Russian, and still is, an alien concept. These propositions may seem self-evident; they describe, indeed, the distinctive characteristics of any society which attempts to elevate the intellectual bankruptcy and moral squalor of Marxism into a political system. Yet it is necessary to insist upon them because the Soviet Union openly declares its desire to impose this system on the rest of the world, and has consistently demonstrated that it is prepared, if it seems necessary and feasible, to use force to do so.

It may come as something of a surprise to a generation brought up on a diet of cynical and carefully orchestrated disinformation to learn that the United States is somewhat different. Its political system is constructed upon a liberal compromise between the demands of freedom and order; its government is open and its press sometimes embarrassingly independent. Its society has none of the residual class preoccupations of many Western European countries, and although its people are refreshingly non-deferential, there is a broad and powerful base of national pride and patriotism. One of the constant surprises for the European visitor is the sight of 30,000 people in a football or tennis stadium standing in silence for the national anthem – a spectacle which would attract the derisive contempt of the rabble which turns European football grounds into battlegrounds, and which greets national anthems with jeers and obscenities.

There is no evidence that the United States plans to use military force to expand its influence or to impose its political system upon the rest of the world. The military alliance of which it is the leader and the central power is entirely defensive – a proposition demonstrable by its nature, its strategic doctrines and its deployment. Simply to speak these sentences is to be conscious that they are statements of the obvious; yet it is clearly necessary to go on saying them as long as there are people in the free world who are at best unable to make moral choices between the two conflicting systems, and at worst disposed to vilify and blackguard the country upon which we depend ultimately for the preservation of our freedom.

For that is the simple fact. If the Western Alliance continues to disintegrate, and if the United States retreats into a carapace of 'global unilateralism', withdrawing its military presence from Europe and engaging in a bilateral relationship with the Soviet Union, the countries of Western Europe will have some hard choices to make, and they are not difficult to identify. One of them has recently been given some airing in a pamphlet written by Mr Robert Jackson, a Conservative member of the European Parliament. In it he writes, '. . . with America on the ebb, Britain faces once again . . . the fact that her central interests are in Europe. A Europe which is forming an identity that in many respects runs counter to American views and short-term interests.'

Mr Jackson's description of America 'on the ebb' reads strangely against the background of the consensus of informed economic opinion, which, on the evidence of most significant indicators, forecasts a more favourable situation in the 1980s for the United States than for most Western European countries. He is, however, the spokesman for a point of view which is being heard increasingly throughout the European Community, and not only on the left of the political spectrum. Its dream is of some kind of *Festung Europa*, a Europe built in conflict with the United States, but in some mysterious way compatible with the wider community of the West. The concept is disturbingly fragile.

There has been no evidence so far that Western Europe is capable even of harmonizing and co-ordinating its foreign policies, much less of forming any kind of political grouping capable of ensuring its own security against military attack.

There is, of course, another option open to the countries of Western Europe. It is, put in its crudest form, to exchange dependence on the United States for dependence on the Soviet Union – for that, as the Finnish position demonstrates, is one of the risks of neutrality. For those convinced of the pacific intentions of the Soviet Union this course holds no terrors; but there should be no doubt that if the North Atlantic Treaty Organization disintegrates into a collection of 'neutralist' nation states, denuded of credible deterrence or effective defence (and that is the logical conclusion of the policies advocated by the 'peace' movements), their survival as free and independent societies will depend upon the whim of the Soviet Union. It is a course which has little attraction for those who are disposed to take the words and actions of the Russian leadership at their face value.

There is, in effect, no real choice for Western Europe. The inescapable necessity is to bend the efforts of foreign policy to repairing the cracks which are appearing in the Western Alliance; to recognize that there are long-term strategic concerns which override short- and medium-term conflicts of economic interests, because they are matters not simply of stability and prosperity but of survival. One of the cardinal aims of Western European foreign policy should be to ensure that the United States remains fully engaged in the security of the free world. An essential prerequisite is to counter the insidious anti-Americanism which is, to the delight of our enemies, beginning to poison the mainstream of the Western Alliance. It is against this background that the governments of the West should be considering their approach to the Strategic Defence Initiative.

3

The Strategic Dimension

It is impossible to isolate the debate about the Strategic Defence Initiative from the general concern about nuclear strategy and the fear of nuclear war, and the political attitudes which these concerns generate. Scientists, historians and other academic figures of impressive intellectual pedigree are constrained, like the rest of us, to grope tentatively toward some kind of truth about one of the most perplexing of the moral problems which afflict the human condition. Some have even been obliged to abandon the reassuring belief that problems must by definition have solutions. Indeed the only people who seem to be secure in their monopoly of wisdom and insight on this matter are those who subscribe to the pure doctrine of disarmament. The only way to achieve disarmament, they assert with an air of profound wisdom, is to disarm ; and if nuclear weapons are the most dangerous and destructive, then they should be disposed of first. The truly appalling difficulties that lie in the way of realizing these simple aspirations are discounted or dismissed as the conspiracy of some vague, sinister and malevolent establishment.

An even more radical formulation of this approach, especially popular in Britain, is the advocacy of unilateral nuclear disarmament. In its most uncomplicated form the argument is that the United Kingdom should in some way renounce all nuclear weapons as potential instruments of war. Some unilateralists seem to believe that other nations possessing these weapons will then see the error of their ways and follow our example. Others believe that even if they do not go as far as that, they will at least recognize our gesture to the extent of excluding us from their targeting arrangements in any future war.

This is, of course, not a new phenomenon. The Campaign for Nuclear Disarmament has been advocating the policy with varying degrees of enthusiasm for twenty-five years or more. In the early 1960s the campaign entered a period of suspended animation, its supporters evidently believing that Harold Wilson's promise to 'renegotiate' the Nassau Agreement with the United States, under which Britain had acquired its fleet of submarine-launched nuclear missiles, really meant that unilateral nuclear disarmament was at hand. The new awakening of the movement in the late 1970s can be attributed to a series of policy statements by the Conservative government confirming the growing suspicion that this was never a serious possibility. For example, the decisions to embark upon a new generation of weapons, the Trident system, to replace the Polaris missiles acquired under the Nassau Agreement; to allow American cruise missiles to be based in the United Kingdom; and to allocate a modestly increased budget to civil defence arrangements have all combined to mobilize the unilateralists again. Although a few new faces have appeared in the ranks, notably that of E.P. Thompson, the battle hymns have a familiar sound. Britain is to have nothing to do with nuclear weapons – no Trident, no cruise missiles, no American bases. 'Ban the Bomb' has come back with a few added grace notes.

It is important that those who hold opposing views about our national security should not dismiss this as yet another artefact from the production line of our thriving protest industry. It is necessary to be aware what kind of people endorse the campaign for unilateral disarmament, and to examine their credentials and their arguments with some care. Before doing so, however, it would be prudent to construct some frame of reference, to articulate some of the assumptions upon which any intelligent policies designed to promote national security within a stable international order must be based. A useful point of departure is that perceptively identified by Professor Michael Howard in his reply (*Encounter*, November 1980) to E.P. Thompson's polemical tract *Protest and Survive*. International conflict,

Professor Howard writes, 'is an ineluctable product of diversity of interests, perceptions and cultures . . . armed conflict is immanent in any international system.' In other words, although war can be avoided, or its effects mitigated, by patent and realistic statesmanship, it will not be eliminated from the conduct of international affairs simply because it is demonstrably cruel and destructive. Indeed it is difficult for anyone not engaged in the business of conjuring up dream worlds to conceive of a stable international system which would not ultimately rely in some way upon the sanction of force.

In this context, one of the primary functions of any government, not excluding our own, is to provide effective arrangements for national security, which must obviously include adequate military defence against any external attack. Before this can be done, the threat of any such attack must be identified and evaluated. It seems reasonable to suggest that the principal, indeed at present the only, external threat to the integrity of the United Kingdom and the survival of its political institutions comes from the Soviet Union. It is unnecessary here to catalogue in detail the various aspects of the immense military apparatus which has been constructed by the Russians in the last twenty years. Even allowing for differences of assessment and interpretation, it is a matter of record that the Soviet Union spends a much higher proportion of its natural resources, both gross and per capita, on its military establishment than does any other great power, and that it has demonstrated on more than one occasion its readiness to use its military strength ruthlessly and effectively in pursuit of its political aims.

In an article in the *Washington Quarterly* (Autumn 1979) Michael Howard suggested that this massive military capability was not necessarily evidence of aggressive intent. There were, he considered,

> . . . too many alternative explanations – atavistic Soviet suspicions of the outside world, the growing collusion of Soviet adversaries East and West, the primacy enjoyed by the military in

> Soviet bureaucratic processes, a determination to demonstrate superpower status in the only way open to her, above all an understandable determination that any further conflict will be fought out on the soil of her adversaries rather than her own. . . .

This seems to be pursuing scholarly detachment beyond the call of duty; and when Professor Howard writes that 'Soviet military capability *as such* is no more evidence of aggressive intent than is that of the United States', he is being either uncharacteristically naïve or wilfully provocative. He knows as well as anyone and better than most that in the business of threat analysis, capabilities and intentions are inseparable; indeed it is often possible to make a valid assessment of intentions *only* from a close study of capabilities.

In the first place it is possible to assert as a general proposition that Russian military strength is entirely disproportionate to any possible requirement for the territorial defence of the Soviet Union, especially in the light of the military capabilities of any potential invader. More specifically, any careful examination of the equipment, deployment, tactical doctrines and training methods of the Soviet armed forces seems clearly to establish an aggressive intent. For example, while the annual manoeuvres of the NATO forces in Europe are regularly based on a battle plan involving an early withdrawal in the face of a Soviet attack, followed by a defensive battle and counterattack, Soviet and Warsaw Pact military exercises include no such defensive phase. They are designed to train their formations in rapid advance, assault, river crossings, airborne operations ahead of the main force, and the use of chemical weapons to neutralize defensive positions. It may be, of course, that this can be explained by the determination of the Soviet Union to fight 'on the soil of her adversaries', but it would be unwise to draw too much comfort from such a theory.

Furthermore, the Soviet Union has consistently demonstrated that it is prepared to use its military capabilities aggressively in any situation in which it cannot clearly and effectively be deterred from doing so. It is sometimes argued that this expansionist tendency springs more from pragmatism and opportunism than from any strategic blueprint, and that it is part of a defensive mechanism designed to pre-empt and neutralize the activity of potential aggressors in the West and the Far East. Those who advance this argument apparently discount the evidence of a number of defectors and dissidents from the Soviet Union and Eastern Europe who have provided persuasive, if not entirely conclusive, evidence of the existence of a grand design for global predominance. The most comprehensive documentation was provided by Major-General Sejna, a high official in the Czechoslovak Ministry of Defence, who handed over to British and American intelligence documents describing a 'Warsaw Pact Strategic Plan' setting out the long-term programme for Soviet expansion. The fact that the KGB took the trouble to try to discredit Sejna, by a disinformation operation designed to cast doubt on his moral integrity and his motives for defection, did little to diminish the significance of his information.

The conclusion of any prudent Western statesman must be that the strength of the Soviet Union, in the context of its known doctrines and policies, poses a real and growing threat to Western security. This is not necessarily the threat of a sudden assault by Warsaw Pact forces in Europe, although such a possibility should never be discounted; it may not even be a threat of direct military action at all. The danger is of a more classical kind, deriving from the political significance of military superiority. If the Soviet Union is permitted to acquire both nuclear and conventional military forces, of overwhelming strength and superiority, the mere threat of military action, whether implicit or explicit, might be enough to ensure that it could achieve virtually unlimited political aims without the need to move a single division across a national frontier.

In this context it is possible to dismiss too readily the fears of some American observers that the Soviet Union might soon achieve the capacity for a pre-emptive strike against the United States. The concern of these people is not that the

Russians might be '100 per cent successful' or be able to 'take out every single land-based missile'. No one with even a rudimentary knowledge of offensive and defensive weapons technology would seriously suggest such a thing. What *is* suggested is that the Soviet Union might be able to destroy a large proportion of the American land-based force, using only a small proportion of its own. It would then be able to threaten with its large residual force to destroy American cities if the Americans retaliated with their bombers and submarine-launched missiles. Critics and sceptics are entitled to discount this possibility. Many of those in the United States who canvass it, however, are men of serious purpose and deep experience; and if it is true, as I believe it is, that the perceptions of adversaries in a confrontation of this kind are as important as objective realities, then the perception of Americans surely deserves the same serious attention as that of Russians.

One of the principal deficiencies of a great deal of strategic analysis in the West has traditionally been a failure to appreciate that the Russian and Western strategic doctrines are based upon entirely different assumptions, cultures and habits of mind. This truth was underlined by an authoritative Russian source in *The Penkovsky Papers*, published in 1965.

> One thing must be clearly understood. If someone were to hand to an American general, an English general, and a Soviet general the same set of objective facts and scientific data, with instructions that these facts and data must be accepted as impeccable, and an analysis made and conclusions drawn on the basis of them, it is possible that the American and the Englishman would reach similar conclusions – I don't know. But the Soviet general would arrive at conclusions which would be radically different from the other two. This is because, first of all, he begins from a completely different set of basic premises and preconceived ideas, namely, the Marxian concepts of the structure of society and the course of history. Second, the logical process in his mind is totally unlike that of his Western counterparts, because he uses Marxist dialectics, whereas they

will use some form of deductive reasoning. Third, a different set of moral laws governs and restricts the behaviour of the Soviet. Fourth, the Soviet general's aims will be radically different from those of the American and the Englishman.

In a great deal of Western strategic analysis, such problems as the disarming strike, ballistic missile defence, selective targeting, modernization of theatre nuclear weapons and civil defence are dismissed too glibly on the basis that 'no one can win a nuclear war'. This statement, uttered with profound conviction by distinguished scientists, retired generals and politicians of various persuasions, is promptly exploited by unilateralists, Soviet apologists and the dottier fringes of the disarmament lobby. Yet it derives entirely from Western value judgements, and finds no place in Russian military thinking. A study of the most authoritative strategic writing from the Soviet Union – that of Sokolovsky, Ivanov, Gorshkov, and Kulikov among others – indicates that the concept of fighting and winning a nuclear war is at the heart of Russian military doctrine. For the Soviet Union the only effective nuclear deterrent is one which confers upon them the option of fighting a successful nuclear war if deterrence should fail. One quotation – from Sokolovsky – is enough by way of illustration:

> . . . we conclude that the Soviet Union's armed forces and those of the other socialist countries must be prepared above all to wage a war where both antagonists make use of nuclear weapons. Therefore, the key task of strategic leadership and theory is to determine the correct, completely scientific solution to all the theoretical and practical questions related to the preparation and *conduct* of just such a war. [Author's italics.]

Britain, like any other advanced industrial country in the West, must clearly ensure that it is able to deter the Russians from any course of action that would threaten its own national security. History has so far shown no way of achieving this without the possession of the classic deterrent against military attack – an effective military defence. If the case put forward by the advocates of unilateral nuclear disarmament is to be examined intelligently and dispassionately, it is necessary to understand first how the military defence of a medium-sized industrial nation like Britain is organized. The basic framework is collective security, based upon the assumption that an alliance of like-minded nations is bound to offer a more effective defence than a number of individual states which could be picked off separately by a determined enemy. Britain has chosen therefore to maintain full membership of the North Atlantic Treaty Organization. To this alliance it contributes conventional forces, composed of regular Navy, Army and Air Force units supplemented by a limited reserve. The bulk of these forces are allocated to and fully integrated within the alliance structure. Although they have certain crucial weaknesses of strength, deployment and equipment they constitute an important and effective element of the conventional defences of the alliance.

In addition Britain maintains a nuclear strike force consisting mainly of four Polaris submarines, each capable of striking at targets in the Soviet Union with sixteen long-range missiles equipped with multiple nuclear warheads in the megaton range. Of these four submarines, one is constantly on station at a high degree of combat readiness. These vessels and their missiles are normally assigned to the Supreme Allied Commander Europe, and their targets are determined by him. In a national emergency, however, it would be possible for the British Government unilaterally to resume the sole responsibility both for allocating the targets and for firing the missiles.

This force, sometimes described as the 'independent nuclear deterrent', therefore has two distinct politico–military functions. In the first place it contributes to the overall power of the Allied nuclear arsenal, the principal function of which is to deter the Soviet Union from using its conventional forces to attack Western Europe by posing the implicit threat of an early resort by NATO to the use of nuclear weapons. Therefore, although it can be argued that the American nuclear capability is so enormous that the relatively small British contribution is numerically irrelevant, that is to ignore the important psychological importance to the Alliance of having a nuclear retaliatory capability outside the direct control of the United States. The second function is to deter the Soviet Union from directing a nuclear attack at the United Kingdom by posing the threat of instant *British* retaliation against Russian cities and military installations. These conventional and nuclear forces, if kept at adequate strength and an appropriate state of combat readiness, combine to provide an effective deterrent against Soviet attack or nuclear blackmail.

Those, like E.P. Thompson, who advocate the unilateral renunciation of the nuclear element in this deterrent, must therefore submit themselves to the intellectual discipline of constructing an alternative system of national security – that is to say if they believe that the country should be defended at all. At this stage it is appropriate to consider the underlying motives and arguments of the unilateralist school. For it has to be said that a small minority of those who advocate unilateral disarmament do so because they wish to see this country defenceless against the designs of the Soviet Union. To categorize them as 'Communists' or 'Marxists' is not an especially illuminating exercise. They are, however, incontrovertibly the agents of Soviet influence in our society. They ply their trade in many disguises: as teachers, professors, television producers, civil servants and Members of Parliament; manipulating and exploiting the machinery of democracy with the single aim of destroying it. Through their penetration of the Campaign for Nuclear Disarmament they have brought their influence to bear upon a broad constituency ranging, as Michael Howard put it in a *Washington Quarterly* article, 'from saintly men of penetrating intelligence to mindless fanatics impervious to reasoned argument'.

There are, however, many who advocate unilateral disarmament from a depth of sincere conviction. They believe in the provision of effective arrangements for defence but argue that the British nuclear striking force is at best irrelevant and at worst potentially damaging to national security. The arguments they deploy usually fall into one or more of three broad categories: moral, political and economic.

The moral argument rests upon the proposition that nuclear weapons are so appallingly and indiscriminately destructive in their effect that to use them in war is immoral and therefore incommensurable with any national interest that might theoretically be secured by their use. This belief has impeccable antecedents in the traditional formulation of the doctrine of the just war – more specifically in the concepts of proportionality and discrimination. It is, of course, possible to argue that, in general, a nuclear weapon is no more immoral than any other weapon used to kill people. It is certainly difficult for a reasonable person to draw any useful distinction of this kind between nuclear weapons on the one hand and, on the other, chemical or microbiological agents, the high-explosive and incendiary bombs which were used to destroy Coventry, Dresden and Hamburg, or even the artillery shells which helped to reduce the abbey at Monte Cassino to a heap of rubble.

Nevertheless, it would be perverse to reject the proposition that nuclear weapons are different not merely in scale but in kind from any other weapons of war; and that their use as weapons of mass destruction against civilian populations would be immoral. There remains, however, the question of whether the threat of their use as a deterrent is also immoral. The position of many Christians is that the conditional intent in this case is no different from the action; that a threat to carry out an immoral act is as immoral as the act itself. This seems to me to take insufficient account of another important element in the doctrine of the just war: that concerning the justice of the aim.

To use a simple everyday analogy, there would seem to be a valid moral distinction between the man who threatens violence in defence of his own person or property and the man who does so in the commission of robbery or rape. Similarly, in the doctrine of the just war, as Julian Lider points out to his book *On the Nature of War*, the justification most widely accepted throughout history and in all belief systems has been defence against aggression. The strategy of the Western Alliance involves the threat of nuclear retaliation to deter a potential enemy from attacking the West, either with his own nuclear weapons or with his demonstrably superior conventional forces. The Soviet doctrine envisages that nuclear weapons might be used in *any* military conflict, including one in which their use became necessary to overcome the enemy's defences. In this context the moral objection to nuclear weapons as a deterrent seems to be less persuasive.

The political or strategic arguments for unilateral disarmament rest upon the proposition that the possession of nuclear weapons is intrinsically provocative, and that if Britain were to abandon them we would cease to be a target for nuclear attack. This ignores the inconvenient historical fact that on the only occasion so far on which nuclear weapons have been used in war, they were used against a country which had no capacity to retaliate. In any case, in the British context it is an obscure argument, since that part of the British nuclear strike force which is submarine-based is invulnerable to a pre-emptive strike directed at the United Kingdom. The cosy assumption that unilateral disarmament would in itself provide us with a nuclear sanctuary has very little basis in reality.

The economic arguments are even less convincing. The standard references to the 'crippling cost of nuclear weapons' with the corollary, spoken or implicit, that they are taking bread out of the mouths of the poor, are not easy to reconcile with the fact that the cost of maintaining the Polaris force is £126 million a year – that is to say, one-fifth of one per cent of the gross national product; or, to put it another way, substantially less than the annual budget of the London Borough of Camden. Much of the public misapprehension about the cost of nuclear weapons springs from the published estimates of the capital cost of a new generation of nuclear missiles, the Trident system, with which the Government proposes to replace Polaris. If the economic arguments were directed specifically to this project, they would carry more weight. It is certainly arguable that £5,000 million might be more intelligently spent in the next ten years, if only in improvements to our severely degraded conventional forces. That is, however, a very different matter from the unilateral abandonment of the existing nuclear force. Furthermore, if the nuclear deterrent were to be abandoned, there would be powerful arguments for a substantially improved conventional capability which could not conceivably cost less than the running costs of the Polaris fleet.

This leads logically to a series of questions which it is legitimate to put to those who advocate unilateral nuclear disarmament. Is it proposed, having disposed of our nuclear deterrent, to remain within the North Atlantic Treaty Organization? Those who believe this to be possible have some serious moral and intellectual issues to resolve. The defensive strategy of NATO is based implicitly not only on the use of nuclear weapons, but in certain circumstances on the *first use* of nuclear weapons. There seems to be some confusion in the minds of people who wish to abandon a nuclear strike force designed principally to deter a nuclear attack on the United Kingdom, while remaining members of an alliance which is prepared to use nuclear weapons to defeat a conventional attack. It is difficult to comprehend a moral position which draws the line at being defended by British nuclear weapons but is prepared to be defended by those of the United States.

There are, of course, a number of unilateralists who follow their argument to its logical conclusion, which is withdrawal from the Western Alliance, removal of American bases and the adoption of a position of neutrality. They, in turn, have some hard questions to answer, this time of a more political and economic kind. If the stance is to be one of unarmed neutrality, it is unlikely that we should survive long as the only industrially developed country in the world without military defences. If, as seems more plausible, the neutrality is to be similar to that of Sweden, Switzerland or Yugoslavia, a very substantial defence budget will be required. As some kind of yardstick it might be worth noting that while Britain in 1982 devoted 11.4 per cent of its government expenditure to defence, Switzerland's proportion was 21.4 per cent and Yugoslavia's 61.6 per cent. Although, for obvious reasons, these figures are not directly comparable, they underline the fact that the renunciation of nuclear weapons does not necessarily lead to a greatly reduced expenditure on defence. Even on the more valid comparison of expenditure per head of population, that of Sweden, the archetypal neutral, is almost exactly the same as our own.

There remains, however, the argument that unilateral disarmament by Britain would be an example which the other nuclear powers would follow. E.P. Thompson's refinement of this proposition concentrates upon the idea of European nuclear disarmament (END) – a revival of the familiar concept of the European nuclear-free zone. The first thing to be said is that no one has yet produced the smallest scrap of evidence that any other nuclear power would follow Britain's example. Indeed all the evidence of over thirty years of arms control and disarmament negotiations suggests otherwise. It is, of course, entirely possible that the Soviet Union might find E.P. Thompson's END a reasonably attractive proposition. Since no member of the Warsaw Pact has any nuclear weapons, nor would they ever be allowed by the Soviet Union to have any, the withdrawal of *Russian* nuclear weapons from Eastern Europe into metropolitan Russia (whence they could still strike at targets throughout Western Europe) would be a small price to pay for the elimination of British and French nuclear weapons and the withdrawal of American theatre-based systems to the United States.

Leaving aside such bizarre asymmetrical arms control proposals, it is necessary to reiterate yet again the basic argument against nuclear disarmament in isolation. In the extremely unlikely contingency that all nuclear powers, shamed by Britain's spectacular *beau geste*, dismantled their nuclear armouries, the military balance would at once be shifted towards the country or alliance possessing the greater conventional strength. It would be impossible without massive expenditure and convulsive political and social upheaval for the West to offer any kind of effective defence, and therefore to pose any kind of credible deterrent, against the overwhelming power of the Warsaw Pact. It is for this reason that the West, throughout the long, tedious and frustrating years of disarmament and arms control negotiations, has always adhered to the belief that nuclear disarmament cannot be considered in isolation from general and complete disarmament under international inspection and control. Until that becomes possible – and the possibility seems more remote than ever – it would be dangerously irresponsible to begin throwing away the one element in the equation which restores the balance and neutralizes Russia's massive military establishment.

While, however, nuclear weapons remain the ultimate deterrent to aggression in a dangerous and heavily armed world, the obvious dilemma persists. If deterrence should fail, the destructive power of nuclear weapons is so immense that their use would create unimaginable devastation and consequences which would go far beyond the territories of states directly engaged, and which would persist far beyond the generations immediately involved.

Is there, then, any way of deterring a potential aggressor, armed with nuclear weapons, without relying exclusively upon the threat of punitive retaliation? In order to answer that question it is necessary first to trace briefly the route by which we have arrived at the present doctrine and posture of deterrence.

4

Mutual Assured Destruction – The Suicide Pact

When the United States had a monopoly of nuclear weapons immediately after the end of World War II, nuclear strategy was a relatively simple matter. It was reflected in a policy generally described as 'massive retaliation', meaning that the United States sought to deter a potential enemy – in practical terms the Soviet Union – from mounting a military attack *at any level* on the West by the threat of immediate nuclear response. This, of course, was effective only as long as the Soviet Union had no nuclear weapons with which, in its turn, to threaten a retaliatory strike.[8]

In 1949, however, the Soviet Union had already tested a nuclear bomb; and in 1953, a year before John Foster Dulles, the American Secretary of State, officially articulated the doctrine of massive retaliation, the first Russian thermonuclear device had been exploded. By 1956, therefore, the Soviet Union was able to threaten most European capitals with nuclear attack; and in 1957, the year of Sputnik, it became clear that the Russians were capable of delivering a nuclear weapon on to targets in the United States. The threat of massive retaliation in response to any military attack at once lost its credibility, although it still remained valid against the threat of *nuclear* attack. At the same time, the missile began to replace the bomber as the principal means of carrying nuclear weapons.

Nuclear strategy therefore began to develop at various levels. At the intercontinental level, a position of nuclear stalemate emerged, a balance of terror in which nuclear weapons had become, for practical military purposes, unusable. As weapons of smaller yield and shorter range were developed, the so-called 'battlefield' nuclear weapons provided the focal point for a new strategic concept known as 'flexible response' or 'graduated deterrence'. The thinking behind this was that if the Soviet Union, with its superior conventional forces, launched an attack on Western Europe relying on the nuclear stalemate to protect its territory against massive retaliation, NATO forces would use battlefield or 'tactical' nuclear weapons to destroy the attacking forces, if necessary raising or 'escalating' the level of nuclear response until it became intercontinental or 'strategic'.

This doctrine was seriously flawed. It relied upon the assumption that the Soviet Union would observe the same measured rules of escalation, being prepared, like poker players, to throw in their hands when the stakes became too high. Russians, however, tend to play chess rather than poker, and there was never any certainty that they would not bring the whole deadly game to a dramatic end by ignoring the carefully constructed ladder of escalation and threatening the territory of the United States itself. As the President of the United States alone could authorize the use of the battlefield nuclear weapons in the first place, there was clearly some doubt as to whether he would put Washington and New York at risk by using nuclear weapons in the early stages of a conventional war. On the other hand, of course, the Soviet Union could never be certain that he would *not*; and thus, by the impact of the element of uncertainty, there was a real sense in which nuclear weapons remained a deterrent against any form of attack by the Soviet Union.

In the 1970s, however, serious doubts were frequently cast on this concept of extended deterrence, in which the United States was assumed to provide a nuclear umbrella not only to protect itself against nuclear attack, but its European allies against conventional assault as well. American observers, notable among them Dr Henry Kissinger, pointed out the dangers of relying indenfinitely on the American guarantee.[9] It was suggested that the European allies should provide more effective conventional defences in order to 'raise the nuclear threshold' – in other words, to increase the delay between the beginning of a conventional attack and the moment at which the United States President would have to take the decision to 'go nuclear'. The expression of these doubts gave rise to fears of 'decoupling' – a situation in which the United States Administration, fearing the effects of a nuclear war, would seek to disengage itself from the military defence of Western Europe.

As one of the means of averting this, Western European leaders, and specifically Helmut Schmidt, the Chancellor of the Federal Republic of Germany, began to advocate the stationing of American intermediate-range missiles in Europe. The Soviet Union had by this time deployed a force of several hundred SS-20 mobile intermediate-range missiles aimed at military and civilian targets in Europe. Fearing that these posed an especial threat in the context of 'decoupling', it was proposed that a balancing force of ground-launched cruise missiles and Pershing II ballistic missiles should be intalled in bases in Western Europe. The reasoning behind this was that, if the Soviet Union used its SS-20s to support an attack against NATO, the use of the cruise and Pershing missiles in response would openly and unmistakably commit the United States.[10]

The habits of strategic thought which provided the climate for this kind of thinking are those of Mutual Assured Destruction. By the gruesome logic of this theory, anything that might interfere with the ability of the potential enemy to inflict 'assured destruction' is inherently undesirable. It is, therefore, inconsistent with the MAD theory to construct ballistic missile defences, or in any other way to diminish the enemy's striking power; nor is it permissible to develop civil defences which would protect the civilian populations who are the hostages in this deadly balance of power. This theoretical model of the nuclear confrontation is in fact no more than a set of assumptions, a largely unconscious dogma which has dominated most of the public debate and much of the official attitude to nuclear strategy. It was the context for the ABM Treaty and the Strategic Arms Limitation agreement concluded between the United States and the Soviet Union in 1972.

It has also provided the doctrinal base for the development of the United States nuclear strike force, which has taken the form of a series of improvements in the 'second strike' capacity, that is, the American ability to survive a nuclear attack and still inflict unacceptable damage on the cities of the Soviet Union. A mutual suicide pact of this kind might have some validity if both sides subscribed to it. Indeed, the entire MAD theory rests on the assumption that the Soviet Union constructs its strategic doctrines and develops its nuclear weapons in the same way. The fact is that it does not, and never has.

Soviet nuclear strategy is based, not on the assumption that a war might be limited, but on the premise that it will inevitably develop into an exchange of nuclear weapons. In this exchange they are resolved not only to survive, but to prevail: it is a war-fighting and war-winning strategy, not one of simple deterrence. The corollary to this is that the Soviet Union is prepared to defend itself against nuclear attack. It has consistently sought to maintain and improve, at substantial cost, its air defence forces and ballistic missile defences; it has also provided civil defence protection, especially for its leadership, its command and control system and its industrial infrastructure. Furthermore, Russian planners have developed the Soviet offensive missile system to implement their own strategy, rather than the one which advocates of MAD would prefer them to have. In other words they have created an array of nuclear weapons of great flexibility, which might, in certain circumstances, be used for a first strike.[11]

Fears of this first began to emerge in the late 1970s, when American intelligence sources forecast the appearance of a 'window of vulnerability' in the 1980s. The basis of this assessment was that, while the United States had been failing throughout the 1970s to take the decisions needed to modernize its nuclear weapons systems and protect them from attack, the Soviet Union had been increasing the yield, accuracy and penetrative power of its own ICBMs. Many observers believed that, by the end of the 1970s, a situation had already been reached in which, *whatever the United States might do to remedy the matter*, a period of great danger would begin soon after 1985. The scenario was that, by that time, the Soviet Union would have achieved a nuclear strike force which would be capable, in a first strike, of destroying a large proportion of the American land-based ICBM force in its silos, using only a relatively small proportion of its own missiles. At this stage, if the United States were to retaliate with what was left of its ICBMs, the Soviet Union would have a large enough force left to destroy the entire United States. It would therefore be in a position to impose a system of nuclear blackmail on the West.[12]

The 'window of vulnerability' theory was not by any means universally accepted. Critics pointed out that it tended to ignore the deterrent power of the other two elements in the nuclear 'triad' – bombers and submarine-launched missiles. Others discounted out of hand the possibility that any Soviet leader would be mad enough to take such a risk, even with the certainty of destroying most of the United States' land-based ICBMs. The general proposition, however, contributed to the general feeling that the 'balance of terror' was moving rapidly out of equilibrium. This was reflected in substantially increased public fears about the dangers of a nuclear cataclysm and the consequent intensification of the activities of 'peace' groups and other protest movements.

Meanwhile, at a different level, there was a persistent and growing concern about the morality and prudence of a deterrent strategy based upon a declared readiness to incinerate millions of Soviet citizens. The fact that such an action would entail the incineration of millions of our own people added substantially to the general anxiety. It was, of course, assiduously exploited by the Soviet Union, with the result that any attempts by the United States and its allies to repair the imbalance of forces and close the 'window of vulnerability' had to be undertaken in the face of growing public hostility. It was into this uneasy and precarious strategic climate that President Reagan launched his Strategic Defence Initiative.

5

The Technology of Strategic Defence

The first thing to be emphasized, since a great many people still seem to be unaware of it, is that SDI signifies nothing more than a programme of research. There is, as yet, no formal intention to take the crucial step of deploying strategic defence systems in space.

It might be appropriate at this stage to defuse the somewhat glib argument that military research has a built-in momentum of its own, which removes it from the processes of rational decision making and leads inexorably to development and deployment. This is simply not so. For example, the enhanced radiation warhead, or neutron bomb, was researched, developed and approved by the NATO governments; but a political decision was taken by the Carter administration not to deploy it. Successive British governments have taken the decision not to equip their Armed Forces with chemical weapons, despite the substantial stockpiles maintained by the Soviet Union, and the preparedness of its troops to use them in the event of conflict. There is an international convention prohibiting the development of bacteriological and toxin weapons, although these have been the subject of long and comprehensive research. Furthermore, the President of the United States, as one of the four points of agreement reached with the British Prime Minister at Camp David in December 1984, undertook that there would be no deployment of defensive systems without prior negotiation.[13]

Much of the argument about the relentless, uncontrollable nature of military research and development derives from the old cliché regarding the 'military-industrial complex'. As there are sixteen major US defence contractors currently involved in SDI research, it is apparently assumed that their greed and arrogance will dictate a final decision to develop and deploy a strategic defence system. It is an attractive theory to those with a built-in suspicion of the capitalist ethic, but it has little foundation in fact. It is true that in the past certain missile systems were developed before there was any clear idea of their strategic value; but on this occasion the President has ordered a programme of research precisely to determine whether or not to develop and deploy strategic defence systems.

Given the major strategic, political and economic implications of a decision to proceed with the deployment of such systems, rigorous standards will have to be met at the research stage. As Paul Nitze, now Special Advisor to the US President at the latest round of negotiations in Geneva has affirmed,

> If the new technologies cannot meet the standards we have set, and thus do not contribute to enhancing stability, we would not deploy them. In that event we would have to continue to base deterrence largely on the ultimate threat of nuclear retaliation, though hopefully at lower levels. However, we have high expectations that scientific and technical communities can respond to the challenge.[14]

The programme of strategic defence research, described in simple terms, is an attempt to discover whether it is technically possible and practicable to devise a defence against nuclear ballistic missile systems which will reduce the presence dependence on nuclear retaliation as the principal means of deterring a potential aggressor. It is not, whatever might have been said or written at the time of President Reagan's original speech, an attempt to create a leakproof anti-nuclear carapace over the United States which would instantly render Soviet missiles useless and obsolete. Fred S. Hoffman, who directed one of the original feasibility studies on the SDI, set out the arguments clearly in his statement before the Senate Armed Services Committee in March 1985:

> If we adopt the MAD view of the role and utility of defenses, and require essentially leakproof defenses or nothing then we will conduct the SDI on what has been called the 'long pole' approach. We will seek first to erect the 'long pole in the tent', that is, we will devote our resources to working on those technical problems that are hardest, riskiest, and that will take longest and we will delay working on those things that are closest to availability. The objective of this approach will be to produce a 'fully effective' multi-layered system or nothing. Unfortunately such an approach increases the likelihood that we will in fact produce nothing and it is certain that it delays the date of useful results into the distant future.
>
> If instead, we believe that defences of moderate levels of capability can be useful then we will conduct SDI in a fashion that seeks to identify what Secretary Weinberger has called 'transitional' deployment options. These may be relatively near-term technological opportunities, perhaps based on single layers of defences or on relatively early versions of technologies that can be the basis for later growth in system capability. Or if they are effective and cheap enough they might serve for a limited lifetime against early versions of the Soviet threat while the SDI technology program continues to work on staying abreast of qualitative changes in the threat. Such an approach would incorporate a process for evaluating the transitional deployment options in terms of their effectiveness, their robustness against realistic countermeasures, their ability to survive direct attack on themselves, their cost, and their compatibility with our long-term strategic goals. Such an approach represents the best prospect for moving toward the vital goals enunciated by President Reagan.

Nor should SDI be confused with the concept of anti-satellite systems (ASAT). Only the Soviet Union has a fully ground-tested anti-satellite system, although the United States is developing an ASAT weapon based on a miniature homing vehicle launched from an F-15 aircraft. Anti-satellite systems were not, however, the concern of the Strategic Defence Initiative, which was designed to provide something which does not at present exist, except in the ABM system deployed around Moscow, namely a defence against attack by ballistic missiles. There is, of course, the problem that most space weapons designed for ballistic missile defence could also be used to destroy satellites. This has obvious implications in the field of arms control.

The theory of the Strategic Defence Initiative is that if it were possible to devise a defensive system capable of destroying a substantial proportion of incoming ballistic missiles (say 50 per cent or more) this would have a significant effect on the nature of deterrence.[15] As a potential aggressor would have no way of knowing *which* five out of every ten missiles launched would fail to reach their target, his options would be limited and his expectations of a successful nuclear strike correspondingly diminished. The argument that if even five missiles out of every ten reached their targets the devastation would still be unacceptable rests on the assumption that what is being contemplated is some kind of war-fighting strategy. It is not. As one feasibility study, commissioned by the US President, acknowledged, even a slight leakage rate of around 6 per cent might be sufficient to create catastrophic damage. This can, however, be discounted in practice.

If an attacker should contemplate a nuclear strike against a large military target, such as a command, control and communications system, as part of a surprise attack, an expected success rate of 6 per cent would be totally inadequate. The more specific the targets, and the higher the confidence of success demanded, the greater would be the relative effectiveness of the defence. For example, in the face of an effective defence system, an attacker might need to aim at least thirty warheads at a single target, since he would have no way of determining in advance the pattern of the defence. Indeed, if he needed to be absolutely certain of destroying the specific target at which he was aiming, he might need to use up to 100 warheads. Clearly, therefore, an attacking force which could be guaranteed to destroy a large military target system *in the absence of missile defences* would be totally inadequate to guarantee the destruction of even a large proportion of the system if it were defended. Yet the expectation of success is essential to the concept of a 'first strike' or surprise nuclear attack.

The argument is reinforced in the case of limited attacks on specific targets, such as military installations in a theatre of operations. If, for example, an enemy were contemplating the possibility of selective nuclear strikes in Western Europe, hoping to do so without the risk of escalation, missile defences would almost certainly rule out the possibility of success, since the number of missiles required would be inconsistent with the aim of containing the level of force. The aim of strategic defence, therefore, is not to 'win' nuclear wars but to maximize deterrence with much less emphasis on the threat of nuclear retaliation.

It must, however, be emphasized that SDI does not imply the complete abandonment of a second-strike, retaliatory nuclear capacity. It *does* imply the ability to maintain it at a much lower level of nuclear weapons.

The concept of strategic defence on which the principal research effort is being concentrated at present is the 'layered' defence – a system designed to destroy enemy missiles at various stages in their trajectory. The typical trajectory of a ballistic missile consists of four phases. In the *boost phase*, which for an ICBM of the present generation lasts about three minutes, the missile is lifted out of its silo and carried through and out of the atmosphere by its first-, second- and third-stage booster rockets. Each rocket burns for about a minute, propelling the vehicle at an increasing speed to a height of about 125 miles. By the end of this phase, the missile is travelling at about seven kilometres a second. The vehicle then enters the *post-boost* (or *busing*) *phase*, which lasts about seven minutes. At this stage, still powered by a low-thrust rocket, it drops off up to ten multiple independently targeted re-entry vehicles (MIRVs) in a programmed sequence and each on its separate trajectory. Along with these the post-boost vehicle may also deploy with each MIRV a number of decoys and other 'penetration aids'.

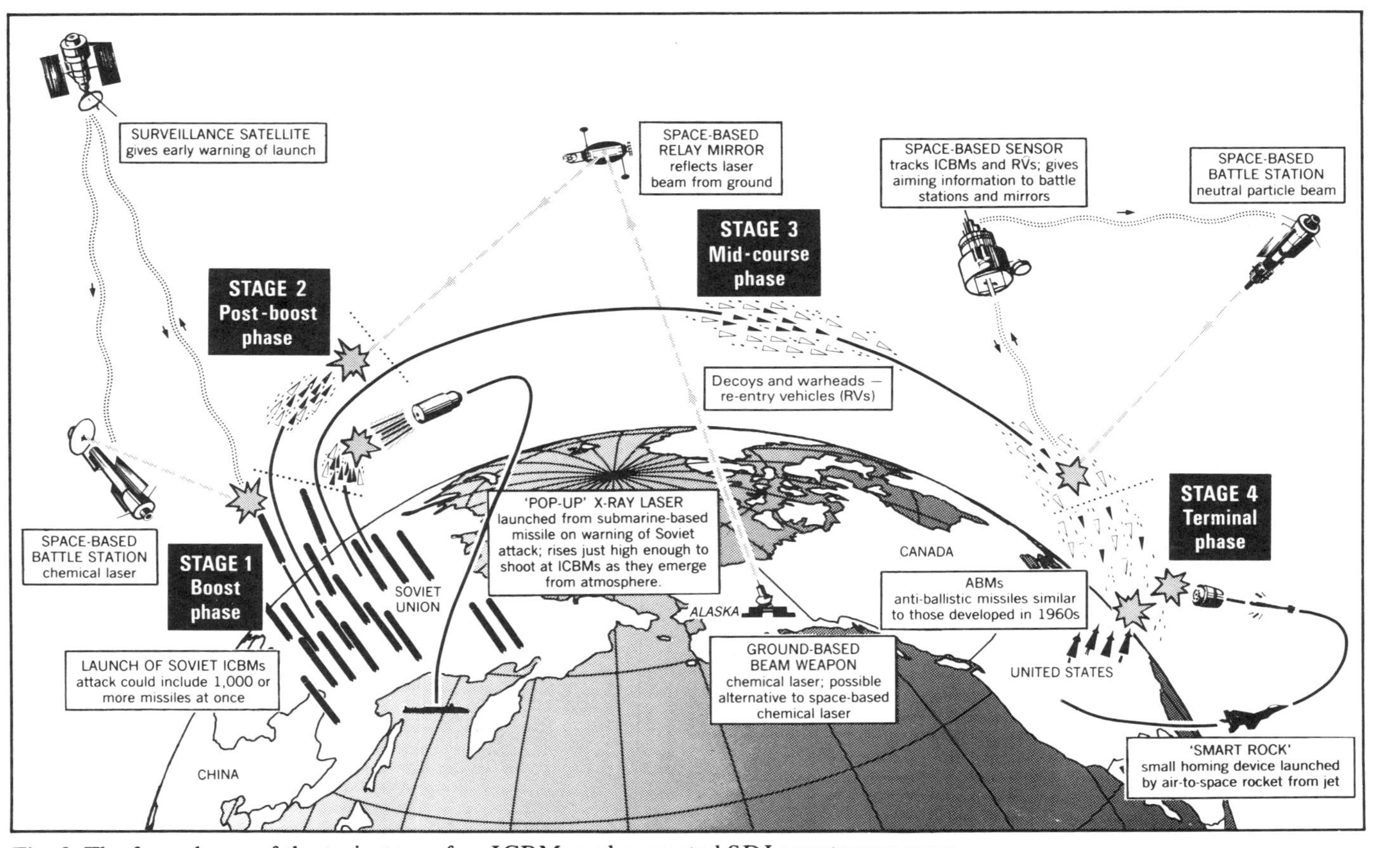

Fig. 2 The four phases of the trajectory of an ICBM, and suggested SDI countermeasures

The MIRVs and decoys then enter the *mid-course phase*, which is entirely ballistic, that is to say, travelling by their own momentum like a bullet fired from a gun. They rise to their highest point or *apogee* at a height of about 625 miles before falling back to earth. This phase lasts about twenty minutes, leading to the *terminal phase*, when the MIRVs and decoys re-enter the earth's atmosphere at an altitude of about 30 miles. After about one and a half minutes of re-entry the warheads, by this time glowing red from the atmospheric friction, detonate over their targets. If it were possible to achieve even a 50 per cent destruction rate in each phase or layer, a simple mathematical calculation shows that only six or seven warheads out of every 100 launched would reach their targets – a 93 to 94% effective defence.

The technology of intercepting warheads in the terminal phase is reasonably familiar ground. It relies on well researched ballistic missile defence techniques and envisages largely ground-based or air-launched systems. The crucial problem is the boost phase. The largest Soviet missiles carry at least ten warheads each, and it is therefore essential, if the calculus of the 'layered defence' is to be achieved, to destroy these missiles at the beginning of their flight trajectory, before the warheads can be separated. But this is also the most difficult problem technically – the problem of destroying a missile deep in enemy territory within seconds of its launch. There are a number of possible techniques. One which is widely discussed at present is the laser, a narrow and intense beam of light or other electromagnetic radiation which, travelling at the speed of light, has virtually what is known as 'zero time to target'. Its effect is to burn through or melt the metal skin of a missile, causing it to disintegrate. Another technique envisages the use of a neutral particle beam, which involves firing a stream of hydrogen atoms, travelling at about 60,000 miles per second. These pass through the skin of the missile and disrupt its computerized guidance system. It is also theoretically possible to intercept missiles in flight with non-explosive pellets or other metal fragments which, on impact, destroy the missile by kinetic energy in the same way

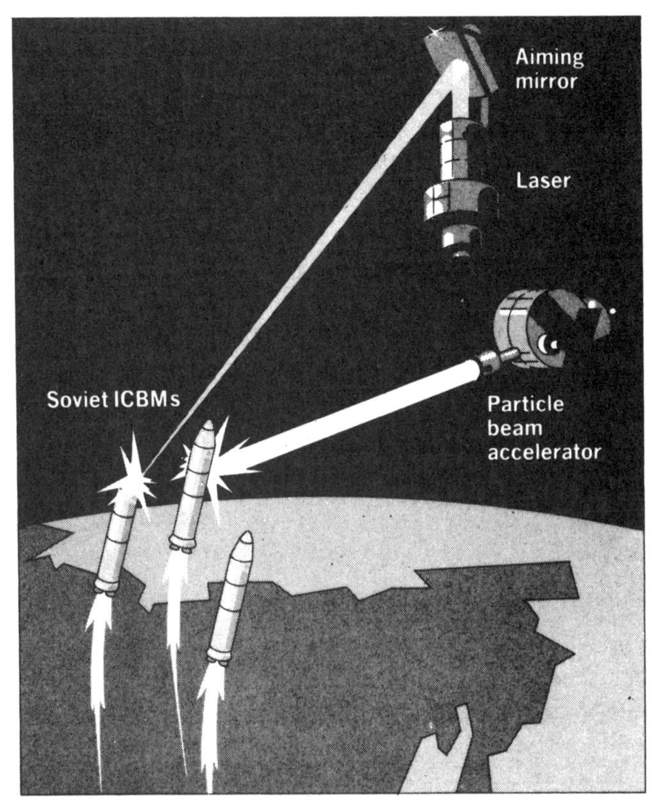

Fig. 3 Space-based beam weapons

These travel in low orbits, so that they sweep rapidly over the earth's surface. They might be chemical lasers, or particle accelerators firing a beam of protons, electrons or whole, uncharged atoms. Atoms are more difficult to accelerate to the necessary velocity, but are more accurate since protons or electrons, which have an electric charge, are for that reason deflected by the earth's magnetic field, while an uncharged, 'neutral' beam is unaffected.

as a bullet. One of the technologies capable of achieving this has been described as a kind of Gatling gun capable of firing a million pellets in one second, with a muzzle velocity of 4,000 feet per second. These form a cloud of pellets 4,000 feet long and hundreds of feet in diameter through which the MIRVs

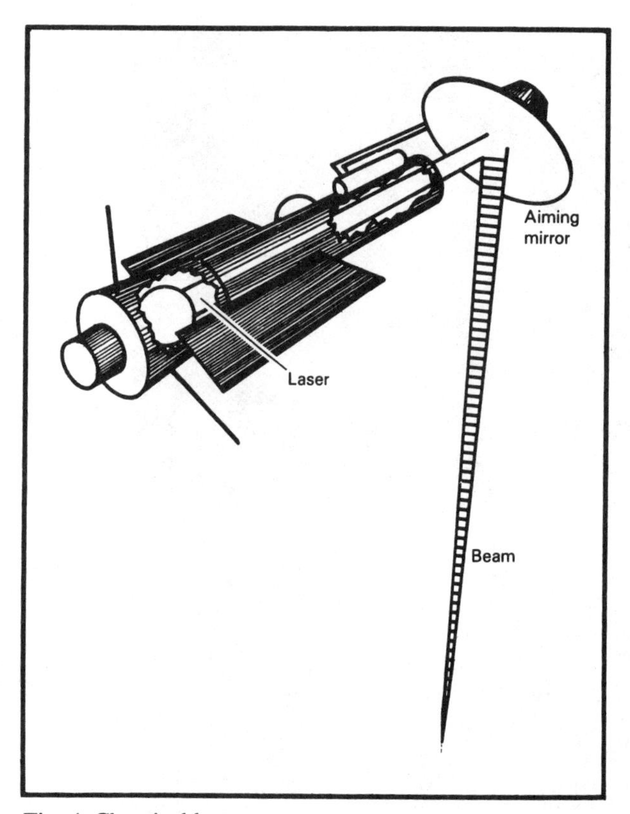

Fig. 4 Chemical laser

When certain chemicals react, they emit infra-red radiation. This can be passed to a laser, which amplifies the radiation into a narrow, intense beam which is then reflected and focused by a mirror that swivels to direct it on to the target. A chemical laser can produce much more power for a given weight than a conventional electrically powered laser: an important consideration for a device mounted on a satellite.

of an offensive missile would have to pass. The tracking and accuracy problems involved in all these technologies are formidable but, according to some US scientists, not insuperable.

The real problem is that, unlike defence systems designed to intercept missiles in mid-course or terminal phases, lasers

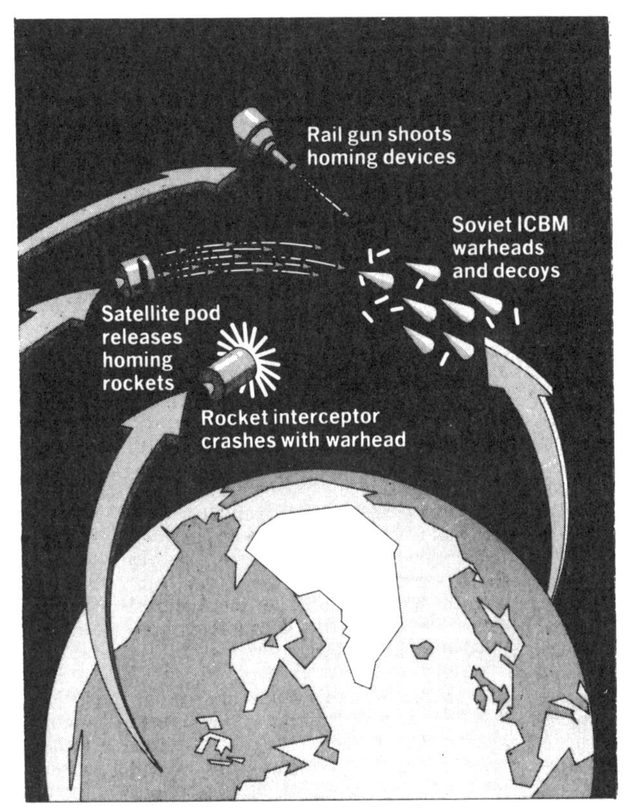

Fig. 5 Kinetic energy weapons

Three alternative types are shown here. All are non-explosive, and destroy warheads by simple impact. The rail gun uses an electric current to fire projectiles along a rail which directs them in the same way as a gun barrel. Otherwise, many small rockets may be fired. These are 'smart rocks', so called because they carry homing devices but no warheads. The third alternative is just to collide a rocket with the warhead.

and neutral particle beams need 'line of sight' basing: that is to say, the generator of the beams must have an unobstructed field of fire directly at all enemy missile launchers. In other words, they, or mirrors to reflect them, have to be based in space. An effective boost-phase defence therefore needs

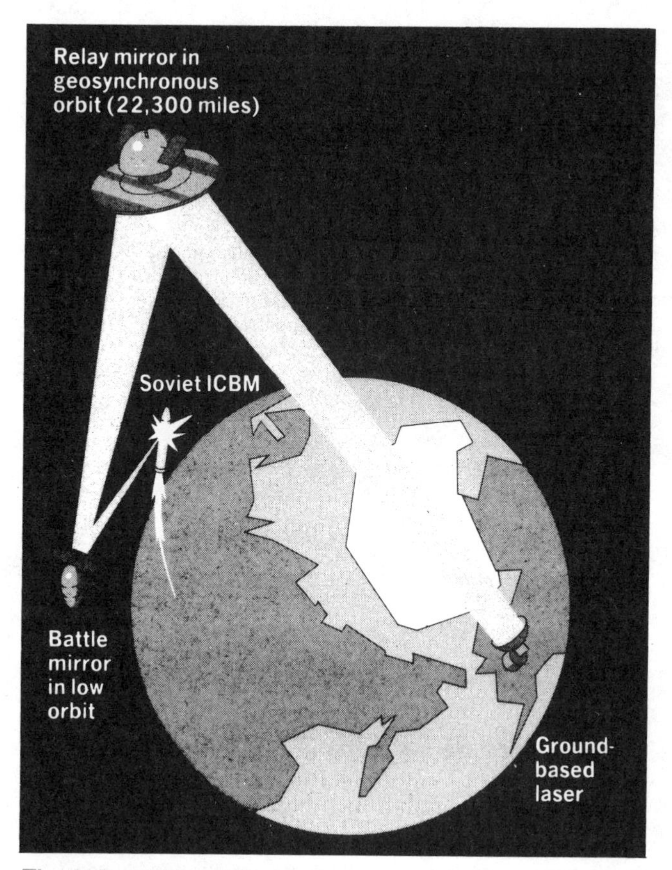

Fig. 6 Ground-based laser

The advantage of installing a laser on the ground is that it can be as heavy as is necessary to deliver any requisite amount of power. Here it is shown directing its beam first to a mirror in geosynchronous orbit, 'hanging' effectively motionless over the equator, and from there to a moving mirror in low orbit which reflects the beam on to the Soviet ICBM.

ideally a fleet of satellites in orbit, large enough to ensure that enough satellites are over the twenty-two Soviet missile fields at any time to attack all 1,400 Soviet land-based ICBMs if, in the worst case, they were launched simultaneously.

An alternative is a 'pop-up' system, which would be de-

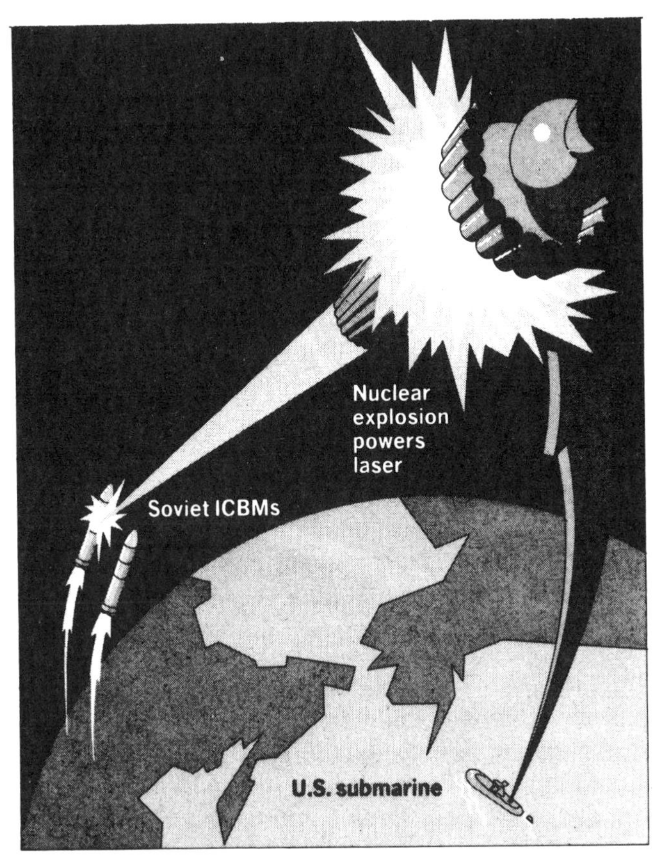

Fig. 7 Pop-up defence

This is rapidly deployed as soon as a hostile ICBM is fired. The advantage is that there is no satellite that could be shot down by the enemy before firing the ICBM; the device is stored aboard a relatively invulnerable missile submarine. The disadvantage is that only a fairly small, light weapon can be launched. The most suitable device so far discovered for the purpose is an X-ray laser.

ployed only at the time of attack – a solution which would reduce the vulnerability of the defences to counterattack, but which would, of course, suffer from formidable time constraints. 'Pop-up' interceptors would be launched from submarines cruising in the Arabian or Norwegian areas. An interceptor of this kind would have to rise to over 600 miles before it could 'see' a Soviet ICBM in its boost phase. This technology, of course, depends for its potential effectiveness on the assumption that the Soviet Union will not build ICBMs with a boost phase so short that no 'pop-up' system could 'see' the missile before the booster had burned out.

The technologies required to achieve an effective layered defence were examined by a panel headed by James C. Fletcher, of the University of Pittsburgh, who directed one of the two studies – the Defensive System Technology Study – commissioned by President Reagan after his March 1983 speech. The other study, on the political implications, was directed by Fred S. Hoffman, who concluded that

> . . . technologies for ballistic missile defenses, together with those for precise, effective and discriminate nuclear and non-nuclear offensive systems are advancing rapidly. They can present opportunities for resisting aggression and deterring conflict that are safer and more humane than exclusive reliance on the threat of nuclear retaliation. . . . The new technologies offer the possibility of a multilayered defense system able to intercept offensive missiles in each phase of their trajectories. In the long term such systems might provide a nearly leakproof defense against large ballistic missile attacks. However, their components vary substantially in technical risk, development lead time and cost, and in the policy issues they raise. Consequently, partial systems – or systems with more modest technical goals – may be feasible earlier than the full systems.[16]

Dr Fletcher, who is a professor of technology and energy resources, assembled a panel of specialists from the armed forces, the defence industries and the universities to examine such aspects of the problem as systems integration, surveillance, target acquisition and tracking, directed energy weapons, battle management and possible Soviet countermeasures. Over a period of four and a half months, the team of fifty scientists and engineers studied the emerging technologies relevant to ballistic missile defence, and concluded that, although substantial problems remain, 'the technological advances of the past two decades show great promise for ballistic missile defence'. Dr Fletcher has described these technologies in some detail.

A ballistic missile defence system has to perform certain essential functions in each phase of the layered defence. First, it has to carry out the immensely complicated task of maintaining a constant watch over the entire enemy ICBM force (*surveillance*); it has to react immediately to the launch of an offensive missile, and instantly compute its trajectory and probable target (*acquisition*). It must then distinguish in the post-boost phase between a warhead and a decoy (*discrimination*). Next it has to monitor the exact trajectory of the missile and its warheads at every second of their flight (*pointing and tracking*); finally it has to direct one of a number of defensive weapons to destroy the missile or its MIRVs (*interception and destruction*). All these activities, as well as accurate assessments of the number of targets destroyed, requiring high-speed data processing and advanced information technology, have to be co-ordinated with infallible accuracy (*battle management*).[17]

One possible conceptual design for a defence system capable of this degree of sophistication begins with satellites in geosynchronous orbit – that is at a height of 22,500 miles above the equator, where their orbital velocity keeps them over a constant point on the earth's surface. They would carry out their surveillance role by means of infra-red sensors capable of detecting an ICBM in flight within seconds of launch, and computers programmed to calculate the general target areas. This information would then be communicated instantly to weapon platforms on up to 100 satellites in lower earth orbit at about 125 miles, and simultaneously to a fleet of mid-course sensor satellites, in orbits from 3,125 to 15,625 miles. These sensors would monitor the deployment of MIRVs and decoys by any missiles which survived the first or boost-phase defence layer.

The boost-phase weapon platforms meanwhile would deploy 'hypervelocity guns' using electromagnetic energy to fire high-speed projectiles on a collision course with the missiles. The kinetic energy released on impact would destroy the missile before it could complete its ascent.

Once the three stages of the booster rocket have burnt out on any surviving missiles, they can no longer be detected by the high-orbit infra-red sensors. At this stage the smaller heat source of the post-boost phase would be picked up by the mid-course sensors, and the missiles would once again be attacked by the hypervelocity guns on the boost-phase satellite platforms.

The mid-course sensors now begin to employ an increasing range of devices to discriminate between MIRVs and decoys, including radar, optical and infra-red sensors. Once the real warheads have been identified, signals transmitted from the space-based sensors guide thousands of small ground-based rockets into the path of the MIRVs. As they approach the re-entry vehicles they release their own warheads, non-nuclear projectiles which home on to their targets and destroy them on impact.

Finally, information from the mid-course sensors is 'handed over' to infra-red sensors carried in high-altitude aircraft, launched on warning of attack. These work in conjunction with radars on the ground to detect any warheads that have escaped the earlier defensive layers, and when the final trajectory is precisely computed, terminal interceptors are launched. As it is necessary to intercept the warheads while they are still high in the atmosphere in order to minimize the effect on the ground of any nuclear explosion, these would be high-acceleration rockets with on-board guidance systems. As soon as they arrive near their targets they explode a cloud of metal pellets into the paths of the descending warhead, or guide a mini-missile, or 'smart rock', on a collision course, destroying the warhead by kinetic energy. This technique of terminal defence was, incidentally, successfully tested on 10 June 1984, when the US Army intercepted and destroyed an oncoming 'enemy' warhead at a height of 100 miles.

Throughout this short, incredibly complicated engagement, a battle management system would operate, consisting of a network of very fast, high-capacity computers in space and on the ground. Each defensive layer also would have its own battle management system, which would direct the engagement in its own layer, and be connected with the systems of other layers to which it could pass on the results of its own intercepts and the details of surviving missiles. The overall C³ (command, control and communications) system would provide the link between all the components of all the layers. One of the major problems to be resolved is the question of whether, once such a system had been programmed and deployed, there would be the need, or indeed the time, for human decision making.[18]

This is, of course, only one model of a possible strategic defence system. The kinetic energy interceptors involved in it are based on existing technologies, but directed energy weapons might in the long term prove to be more effective. 'Directed energy' is a term that has passed into the general vocabulary of the SDI debate, and it is used to describe three kinds of beam weapons based on the laser, radio-frequency (microwave) devices, or particle beams. A laser weapon employs an intense beam or pulse of coherent electromagnetic radiation – that is, with its waves organized in step so that they reinforce each other. It is aimed at a target by an aiming device similar to a telescope. The beam can be of visible light, infra-red or ultra-violet radiation, X-rays or gamma rays, and its effect would be to damage an attacking missile by melting or vaporizing its surface and damaging its internal components. In a radio-frequency weapon, electromagnetic radiation at wavelengths similar to that of radar is aimed at a target by means of an antenna. The principal effect would be to damage or destroy the electronic circuits of a missile. A particle beam weapon produces intense beams of electrons, protons, atoms or ions by means of a high-energy accelerator; these would strike deep into the attacking missile, causing its own materials to produce damaging secondary nuclear radiation or X-rays. Some of these technologies are suitable for use only in space or at the upper fringes of the atmosphere, and not only because they

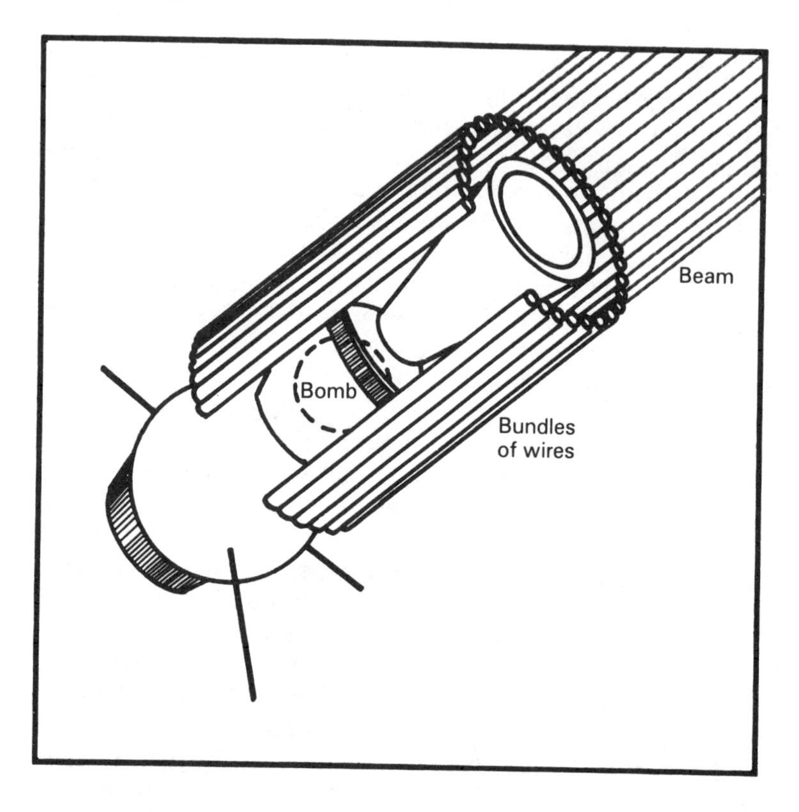

Fig. 8 X-ray laser

This is fired from a submarine, so it has to be fairly small and light. It also has to be powerful, so it is powered by a nuclear bomb. The bomb is surrounded by parallel bundles of metal wires, which are aimed at the target. In the microseconds before the wires are vaporized, radiation causes a vast pulse of X-rays to flow out of their ends.

can only be used in 'line of sight', with the target physically in view. There are, for example, fundamental physical reasons why the X-ray laser is not suitable for use against targets on the ground or in the lower atmosphere since its generation requires a nuclear explosion.

It is obvious, therefore, that if directed energy weapons are to be used for boost-phase interception, they will have to be mounted on satellites, or possibly use beams aimed up from ground stations and reflected on to the target from arrays of mirrors on orbit. Whatever technology is eventually used, even by the most optimistic assessment it would not be possible to construct a multi-layered defence system of this conceptual model today. It is for this reason that the Fletcher Report recommended the programme of research which is now being carried out: precisely to determine whether certain critical technological problems can be solved – in other words, whether an effective ballistic missile defence is indeed possible.

6

Feasibility and Countermeasures

At the heart of the debate about strategic defence is an issue which for the non-specialist is difficult, if not impossible, to resolve: technical feasibility. The scientific debate is predictably opinionated. The temptation for the layman who tries to follow it is to express the hope that he might one day be as sure of *anything* as the scientists, especially those opposed to SDI, seem to be of *everything*. On the one hand James C. Fletcher offers the modest opinion that 'work during the past few years has shown that technological promise is rich, and that effective defenses may indeed be possible'. On the other hand Professor Hans Bethe, a Nobel Prize winner in physics in 1967, has stated that '. . . the technologies required for a defense of our population against nuclear armed missiles are far beyond the state of the art and in most cases are unlikely ever to work effectively.'

Much of the case that strategic defence is not technologically feasible is based on the assumption that in order to be effective it has to provide 100 per cent 'leakproof' defence. This fallacy, possibly deriving from President Reagan's March 1983 speech, reflects a misunderstanding of the nature of research into ballistic missile defence. It is worth repeating that the aim is not to provide 100 per cent protection, either for the population of the United States or its retaliatory missile force. It is to demonstrate a capacity to destroy so many attacking missiles that the Soviet Union would not know how many targets, or which targets, would be destroyed. This would make a first nuclear strike an even more problematical option than it is today, thus increasing the credibility of the deterrent.

It is likely that the 'leakproof' fallacy was at the heart of the first major scientific broadside to be delivered against the Strategic Defence Initiative. In March 1984 a report was published under the sponsorship of the Union of Concerned Scientists, signed by a number of scientists of high reputation and distinction. It concluded, among other reservations against SDI, that 'thousands' of satellites would be needed to provide the defensive screen; that one of the devices under consideration would require a satellite weighing 40,000 tons to be placed in orbit; that the power needed for the lasers and other devices involved would be the equivalent of 60 per cent of the total power output of the United States; and that the whole strategic defence could be easily defeated by relatively simple countermeasures.[19]

At about the same time the Office of Technology Assessment of the US Congress (OTA) published a report which arrived at the conclusion that the chance of protecting the American people from a Soviet missile attack was 'so remote that it should not serve as a basis for public expectations or national policy'. Both these reports, documented with impressive charts and tables and carrying the imprimatur of some of the world's most celebrated scientific names, carried enormous weight. Dr Robert Jastrow, a nuclear physicist and founder of the Goddard Institute for Space Studies, pointed out in an article in *Commentary* (December 1984) that if the UCS calculations were correct there would be no point in looking any further into the SDI plan. Each satellite might cost as much as a billion dollars, and if 2,400 were really needed, the price would clearly be too high even for the United States.

However, as Dr Jastrow comments, computer studies carried out at the Livermore laboratory suggested that 90 satellites might be enough and, if the satellites were in low orbits, as few as 45. In later testimony before a congressional committee, a spokesman of the Union of Concerned Scientists explained that the original estimate of 2,400 had been based upon an incorrect assumption about the pattern of Soviet missile bases, and reduced the estimate to 800; it was subsequently reduced still further to 300; and the report of the

Office of Technology Assessment estimated that the required number of satellites would be 160.

Similarly, in the UCS study, the case for the use of neutral particle beams in space was ostensibly demolished. It would clearly be absurd to contemplate placing in orbit a satellite weighing 40,000 tons. Yet, as Dr Jastrow points out, a spokesman for UCS also corrected this estimate in hearings before the Senate Armed Services Committee, saying that the original calculation was based on the fact that '. . . we proposed to increase the area of the beam and the accelerator, noting that would make the accelerator unacceptably massive for orbital deployment. Our colleagues have pointed out that the area could be increased after the beam leaves the small accelerator.' In fact Dr Jastrow suggests that the correct estimate of the weight of the 'small accelerator' needed to generate the neutral particle beam is about 25 tons.

In the report of the Office of Technology Assessment there appears another calculation which, if correct, would seem to cast doubt on the entire credibility of the Strategic Defence Initiative. In considering the requirements of terminal defence the report based its conclusions on the 'smart' mini-missile technology and concluded that it would require 280,000 such missiles to provide an effective defence, once more casting serious doubt on the cost-benefit balance of the system. Other professional analysts estimate that at most 5,000 interceptor missiles would be needed. The remarkable disparity derives from the basis of the calculations. The OTA report was based on the assumption that a full-scale Soviet attack might be aimed *at any one* of a wide spectrum of targets in the United States and that therefore *every target* would have to be protected by enough interceptors to counter the entire Soviet attack. While Dr Robert Jastrow's description of this kind of calculation as 'GIGO' (the computer programmer's acronym for 'garbage in, garbage out') may sound rather harsh, it certainly seems to be an unreasonably exaggerated use of the 'worst case' scenario.

Wild exaggerations of the technical problems have led to

similarly wild exaggerations of cost. In fact, as General Abrahamson, the Director of the Strategic Defense Initiative Organization (SDIO), testified to the United States Senate in March 1985, it is estimated that SDI would cost about $26 billion between 1985 and 1989. The appropriation for fiscal year 1985 is in fact $1,400 million and the request for fiscal year 1986 just over $3.5 billion – not outrageous sums in the context of an overall defence budget currently running at over $200 billion a year. (The term billion is used in its American sense, denoting one thousand million.)

Further doubts about the technological feasibility of strategic defence have arisen from a consideration of possible Soviet countermeasures. The most obvious of these would be to increase the size and power of the offensive missile force. Indeed, in March 1985, General Chervov of the Soviet General Staff, speaking in an interview about possible deployment of a strategic defence, said specifically, 'We are not going to sit on our hands and wait . . . we'll start to perfect our strategic forces before that time.' While making allowances for an element of minatory rhetoric in this kind of statement, it is necessary to consider seriously the possibility that the Soviet Union might seek to saturate any ballistic missile defence system by increasing the number of its missiles and warheads. The obvious implication – indeed the conclusion arrived at by the Office of Technology Assessment – is that the United States would have to increase the number of its defensive satellites in direct proportion to the increase in the number of Soviet missiles. On the other hand, theoretical physicists at Los Alamos insist that this is an error, and that the required number of satellites increases in proportion to the square root of the number of offensive missiles. The disparity, once again, lies in the basic assumptions on which the calculations are based – in this case assumptions about the spacing of defensive satellites in their orbits. As it is reasonable to believe that if a strategic defence system were ever deployed, the spacing would be designed to optimize the defence, a 'worst-case scenario' in this instance is not merely unreasonable, it is perverse.

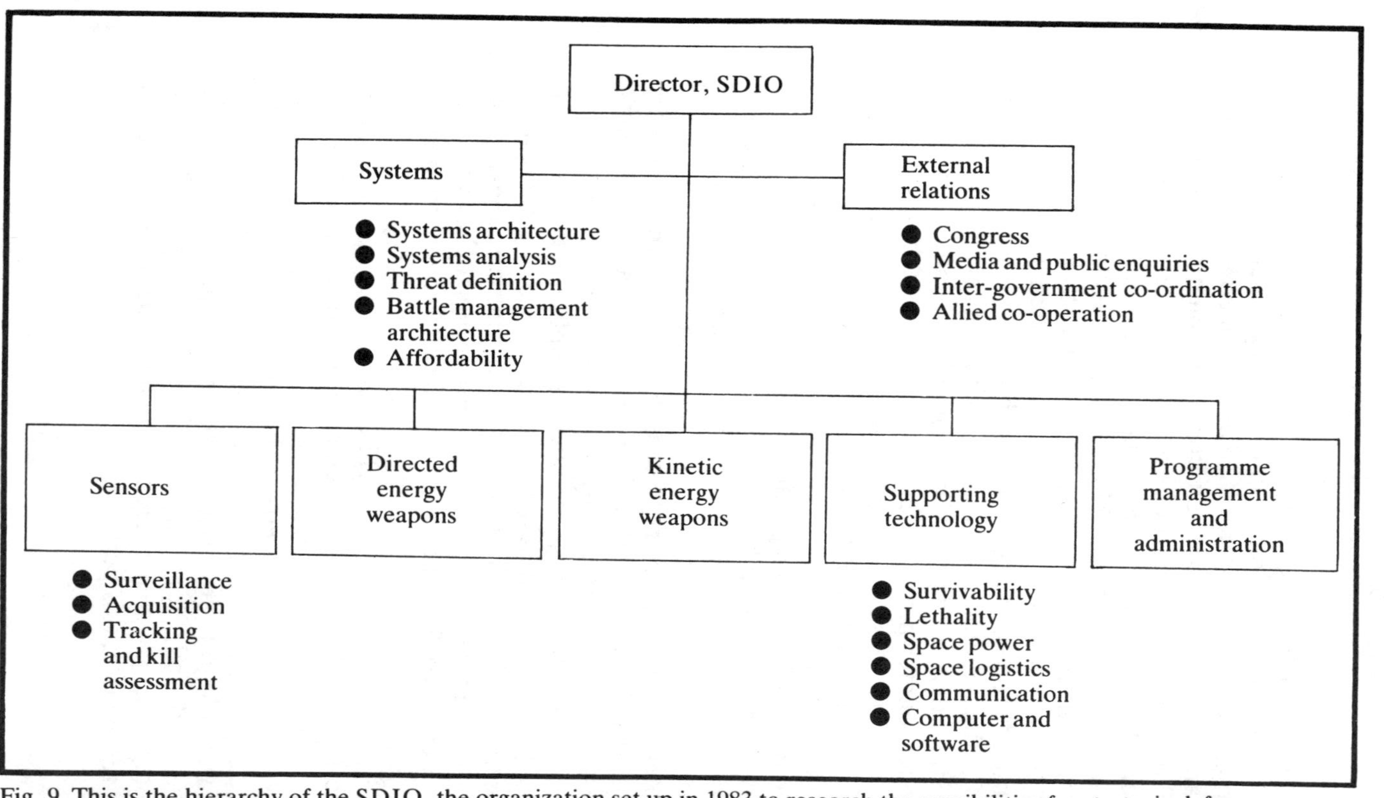

Fig. 9 This is the hierarchy of the SDIO, the organization set up in 1983 to research the possibilities for strategic defence. The main areas of investigation are shown. It is purely a research body; should any system actually be deployed, staff would be quite differently organized.

It is possible, then, to conclude, as Dr Jastrow does, that a Soviet attempt to saturate a strategic defence system would prove to be an expensive and ineffective countermeasure. If it is assumed that the United States originally deployed 100 defensive satellites and that these could destroy 80 per cent of Soviet offensive missiles (a conservative estimate according to the advocates of strategic defence); and if it is assumed, reasonably, that the Soviet Union would seek to nullify these defences and ensure that the number of missiles getting through the defensive screen would be the same as would have been the case if there were *no* defences, they would have to multiply the size of their existing offensive force by five. As they have at present 1,400 missiles, they would have to build another 5,600 missiles and silos, bringing the total up to 7,000. If the Los Alamos square-root rule is then applied, this would mean that the United States could counter this increased threat by deploying fewer than 100 additional satellites. Clearly, on this assumption, the Soviet Union would quickly be bankrupted if it chose to follow this course.

This conclusion, however, assumes that no other countermeasures would be employed. One of the more attractive options for neutralizing a space-based defence system would be to attack the satellites on which it would depend. The Soviet Union has already tested an anti-satellite (ASAT) system, but it depends for its launch on precisely the kind of missiles which would themselves be vulnerable to the interceptor satellites. Any strategic defence system would obviously include protective measures against this kind of counterattack. An alternative means of destroying the defensive satellites would be to use orbiting anti-satellite satellites – a kind of space mine which would move into orbit alongside its target and explode on command. This would, of course, require extremely accurate station-keeping, especially to track an interceptor satellite capable of constant slight changes of orbit. In any case, if the Soviet Union were to contemplate a nuclear attack on the United States, but had to attack and destroy American interceptor satellites before launching it, a substantial degree of the element of surprise would obviously be lost.

Other, more exotic forms of countermeasure have been suggested from time to time. These include protecting offensive missiles from the neutral particle beam with a lead shield; imparting a high shine to the surface of the missile to reflect part of a laser beam and thus diminish its effect; spinning the missile so that the energy of the laser is spread over the whole circumference instead of being concentrated on one spot; fixing to the missile a metal skirt or 'band-aid' which would slide up and down the surface, automatically covering the spot attacked by the laser beam and protecting it from the full heat; and the 'window shade', a flexible metallic sheet to protect against the X-ray laser. The drawbacks to these ideas, as seen by the supporters of SDI, are substantial. A lead shield thick enough to neutralize the neutral particle beam would, it is suggested, have to be at least 1½ inches thick and would therefore weigh several tons – considerably more than the payload of the missile. Polishing the missile would have only a modest and temporary effectiveness against laser attack; and spinning the missile, by definition, only reduces the effect of a laser beam by a factor of pi. 'Band-aids' and 'window shades' fitted to the relatively fragile skin of a missile (the ratio of the weight of a missile loaded to that of one empty being between 10 and 15 to 1) would, it is claimed by missile technologists, require a major redesign of the propulsion system.

Countermeasures against strategic defences in the terminal phase are likely to be most effective. The most familiar of these are known as 'salvage fusing' and 'echeloning'. (Strategic planners and systems engineers are not too concerned with elegant use of the language.) Salvage fusing is the technique of programming a warhead so that it detects when it is about to be destroyed by enemy defences and automatically explodes so as to cause damage to the defences rather than just being neutralized. Echeloning simply means detonating warheads at intervals instead of simultaneously so that the chance of survival of some warheads is increased – a form of dispersal, a concept familiar in more conventional forms of warfare. By a combination of salvage fusing and echeloning it would theoretically be possible to create a continuous thermonuclear chain explosion reaching from 1,000 miles out in space to the target on the ground.

There is no disposition among those involved in the strategic defence research programme to discount the importance of countermeasures. Dr Gerold Yonas, Chief Scientist of the Strategic Defense Initiative Organization, has summed up the problem as follows:

> The ability of any defensive system to survive in the face of direct attack upon it and to continue to function as effectively as possible, even if degraded by attack, is another of the important matters to be resolved. Space-based components that must orbit directly over the Soviet Union will face a host of possible threats, including direct-ascent ASAT weapons, ground- or space-based lasers, space mines, particle beam weapons, and the effects of nuclear explosions. The tactics of survivability are accustomed ones – hardening, active self-defense, concealment, proliferation, maneuvering – but applying them in a cost-effective way to future defenses will be a difficult challenge.

In effect, there is virtually no agreement among scientists of high repute and undoubted integrity regarding the technological feasibility of the Strategic Defence Initiative. Seekers after the truth can do no better than to await the outcome of the research programme, which is designed to resolve some of these divergences of scientific opinion. It is, however, wise to make allowances for the prejudices of scientific theorists. Careful and scrupulously honest as they may be, there is a human tendency to check the results of theoretical calculations less thoroughly if they appear at once to coincide with the expectations of the theorist. In this context many of the critics of the Strategic Defence Initiative are suspicious of the findings of scientists and others who are held to be associated with the armed forces or the defence industries – the familiar 'military-industrial complex'. On the other hand, advocates of strategic defence point out that those who suggest that it is *not* technologically feasible have their own ideological prejudices. Indeed, the Union for Concerned Scientists, or those who wrote its report, stated their belief that a defence against Soviet missiles would 'have a profoundly destabilizing effect on the nuclear balance, increasing the risk of nuclear war' – a highly subjective political judgment which has no direct relevance to the technological feasibility of strategic defence, but might have considerable subconscious effect on the findings of the scientists concerned. Professor Hans Bethe, in reaching his conclusion that SDI could not be technologically effective, went on to say, at a hearing of the Armed Services Committee in November 1983:

> . . . it is also puzzling why anyone should believe that that is the road to a less dangerous world, for a direct, cheap and safe road is known to exist; negotiated and verifiable deep, deep cuts in the strategic offense forces, and non-nuclear alternatives to our excessive reliance on nuclear weapons.

Now, that is not a scientific observation; it is a political and military judgement, and a deeply tendentious one at that. It is by no means self-evident that 'deep cuts' in strategic offensive forces alone are necessarily the road to a less dangerous world; and Professor Bethe seems to ignore the fact that the Strategic Defence Initiative is, precisely, one of the 'non-nuclear alternatives' to which he refers. Dr Lowell Wood, a scientist at the Livermore laboratory, has asked, somewhat rhetorically: 'Is Hans Bethe a good physicist? Yes, he's one of the best alive. Is he a rocket engineer? No. Is he a general? No . . . he dabbles as a military systems analyst.'

However, even if one of the world's great physicists is not necessarily to be regarded as an infallible source of political and strategic wisdom, the Strategic Defence Initiative does raise political and strategic issues, which are at least as important as those of technological feasibility.

7
Soviet Strategic Defence Research

One of the principal political elements in the SDI debate concerns its impact on the strategic balance between the United States and the Soviet Union, and one of the arguments most frequently directed against research into strategic defence is that it will force the Soviet Union into embarking on a similar programme – thus, to use the language of the critics, 'creating another dangerous twist in the spiral of the arms race'. This argument is much reinforced by the public statements of the Soviet leadership, which condemns the American initiative as 'dangerous, destabilizing and provocative'. Furthermore, in 1984 a working group of the Committee of Soviet Scientists for Peace Against the Nuclear Threat published a report which concluded that space-based systems are too expensive, that they are technically unattainable and that they are easily neutralized by countermeasures. Yet, in January 1985 Nikolai Basov of the Soviet Academy of Sciences announced in Moscow that the Soviet Union would have no technical difficulty in matching the American SDI programme.

It may or may not be significant that the arguments of most Western critics of SDI include one or more elements of the Soviet position. On the one hand it may signify a substantial success for Soviet disinformation; on the other, it may simply reflect the community of interest and intellectual integrity shared by peace-loving people on both sides of the great divide. In any case, it is possible to point to a certain internal inconsistency in the Soviet position. Surely, if the SDI is useless, expensive and easily neutralized, it cannot at the same time be dangerous, destabilizing and provocative; and either it is technically unattainable or it is well within the reach of Soviet scientists.

What appears to be a puzzling contradiction in Soviet attitudes can, however, be easily understood if one simple proposition is accepted: namely that the Soviet Union has been engaged for many years, in conditions of characteristic secrecy, on its own programme of research into space-based defence, and that it now fears that it may be overtaken by a similar programme which would have behind it the full weight of Western technology, industrial infrastructure and economic resource. There is, indeed, substantial evidence of this; and it would be strange if it were not so.

The Soviet rejection of the doctrine of Mutual Assured Destruction as 'naïve and bourgeois' is reflected in a great deal of its military writing and evident in its military policies. Russian planners believe that nuclear war is possible; indeed, that if war begins it will inevitably develop into a nuclear war. Their purpose is to survive and to win such a war. In pursuit of this aim the Soviet Union has spent, in the last twenty years, roughly as much on defence as it has on massive offensive forces. It has the world's only operational ballistic missile system protecting Moscow, and it has an operational anti-satellite (ASAT) system. The SA-X-12 surface-to-air missile system, as has been noted, has the potential capability to intercept some types of ballistic missile.

Furthermore, the Soviet Union has constructed large phased-array radars, notably at Krasnoyarsk where the installation is able to provide target tracking data for a ballistic missile defence. In addition to these easily observable signs of a close interest in strategic defence, there is evidence that the Soviet Union has been engaged at least as long as the United States in research into the military applications of directed energy, with special emphasis on laser weapons.

According to recent American intelligence assessments, Russian technology in high-power, high-energy lasers for weapons application is at least as well advanced as that of the United States, and there is evidence of the development in the Soviet Union of prototype laser weapons, including a

Fig. 10 Locations of key Soviet directed energy and ballistic missile defence facilities

ground-based laser with an anti-satellite capability. The USSR could have an operative space-based anti-satellite laser system in the 1980s, and space-based anti-ballistic missile systems by the turn of the century. There has also been impressive progress in particle beam technology and in microwave weapons.

Although Russian research and development programmes have tended to concentrate on the various laser technologies, impressive progress has also been made with particle beams. Research into linear accelerators has obvious applications in normal high-energy physics for peaceful purposes; but it has equally obvious potential in the military field. Similarly, the Soviet Union's expertise in microwave technology is at least as great as that of the United States.

None of this, of course, will come as a surprise to any student of Soviet strategy. Russian strategists have consistently emphasized that their planning is based on the calculation that any conflict between NATO and the Warsaw Pact will inevitably escalate to a nuclear war. In this war, it will be their purpose to win; and they know in any case that effective defence is the essential ingredient of a credible deterrent.

It is for these reasons that they have placed far more emphasis on civil defence than any NATO government. On mobilization, the Soviet Union would have a nationwide civil defence force numbering in the region of 16 million, and a network of hardened deep shelters capable of accomodating at least 175,000 officials, in addition to its local urban shelters.[20]

The strategists have also ensured the protection of Moscow by constructing the only existing operational anti-ballistic missile system: the so-called *Galosh* system, based on the 100 interceptor missiles and six radar complexes permitted under the 1972 ABM Treaty. They have recently upgraded it. It has devoted substantial resources to the creation of hardened command and control centres, and has developed its conventional air defences at a time when the United States has neglected its own. It has some 4,000 interceptor aircraft to the 252 of the United States, 7,000 warning systems – including satellites, early warning and ground-control intercept radars – and 9,600 surface-to-air launchers. The corresponding systems in the United States are either minuscule in comparison or entirely non-existent. The Reagan administration has requested funds for a new network of radar installations, interceptor aircraft and surface-to-air missiles to buttress the ageing American air defence system, most of which dates from the 1950s. Its efforts have so far encountered resistance in Congress.[21]

In addition to large phased-array radar systems, which some Western authorities believe to be in contravention of the ABM Treaty, the Soviet Union has begun to deploy a new surface-to-air missile, the SA-X-12, which is capable of intercepting some types of Western ballistic missile. It would, therefore, have been surprising if Soviet scientists and military planners had not carried out serious research into space-based strategic defence systems. They do not, of course, have the kind of political system in which this has to be the subject of a presidential announcement; nor, if it had been announced, do they have the kind of press which would have been ready or free to apply the 'Star Wars' label in its headlines, and to publish columns of hostile and derisive comment.

Those who are disposed to suggest that the United States should abandon its strategic defence research programme in the interests of arms control negotiations might do well to ponder one significant aspect of the public debate. Within a month of President Reagan's SDI speech of 23 March 1983, a letter had appeared in the *New York Times* attacking the initiative. It was signed by over 200 Soviet scientists – a powerful blast, one might have thought, against the whole idea of strategic defence. What might have escaped general notice at that time, however, was that many of the signatories were Russian scientists who have spent their own professional lives designing Soviet strategic missile systems, military aircraft and nuclear submarines. Possibly the most interesting signature was that of Evgeny Velikhov, Vice-President of the

Soviet Academy of Sciences and the leading figure in strategic defence research in the Soviet Union.[22]

The Soviet Union is clearly already well advanced along this road. The suggestion, therefore, that the American Strategic Defence Initiative might provoke the Soviet Union into a similar programme is considerably wide of the mark. Ballistic missile defence development began in the United States in the late 1950s, when the Nike-Zeus programme was devised to protect the Americans from nuclear attack; by 1969, various changes in technology and strategic thinking had replaced this with the Safeguard programme, designed to protect American missile silos. Meanwhile, in the 1960s the Soviet Union was making a major effort to build up a ballistic missile defence system; ABM sites were constructed around Leningrad and the Minsk Highway. At this time, Soviet strategists were openly stating that anti-satellite and anti-missile systems would become major elements of their National Air Defence forces. In the section on protection from enemy nuclear attacks in the 1963 edition of *Military Strategy*, Marshal Sokolovsky made a statement which should be engraved in large letters on the walls of any conference room in which the Strategic Defence Initiative is being discussed.

> In our country the problem of eliminating rockets in flight has been successfully solved by Soviet science and technology. Thus, the creation of an invulnerable anti-missile defence has become quite possible. [He went on to describe the basic requirements of an anti-missile defence system:] . . . powerful radar (ground and airborne) or other automatic technical equipment (on artificial earth satellites) to assure the long-range detection of missiles during the boost phase . . . the co-ordinates of the missile's flight trajectory . . . electronic countermeasures to assure the deflection of the missile from its intended target and, possibly, to destroy it in its trajectory.

If one of the Soviet Union's leading military could write this over twenty years ago, it is difficult to understand the position of those in the West who now characterize the Strategic

Defence Initiative as 'provocative'. Furthermore, when he visited London in February 1967, the Soviet Prime Minister, Kosygin, asked about ballistic missile defence, said: 'I do not believe that a defensive system intended to forestall attack can be the reason for the arms race. An anti-missile system may be more expensive than an offensive system but its task is not to kill people but to save lives.'

Throughout the last twenty-five years research and development into ballistic missile technology has continued systematically and quietly on both sides. President Reagan's speech of March 1983 was, therefore, not the beginning of a new programme; and even if the Soviet Union genuinely regarded it as something new and provocative, the impact on their own programme need not give rise to any great degree of concern in the West – their strategic defence researches are already as intensive as they could reasonably be. A much more legitimate political concern is in the field of arms control.

8

The SDI and Arms Control

The implications of the Strategic Defence Initiative in the field of arms control fall into two main categories: its relevance to existing arms control agreements, and its possible impact on current and future arms control negotiations.[23] It can be demonstrated with a reasonable degree of certainty that *research* into ballistic missile defence, whether based on the ground, in the atmosphere, or in space, contravenes no existing arms control agreement.

The most significant existing agreement in this context is the Anti-Ballistic Missile Treaty of 1972 (Appendix 2). This treaty forbade the development by the United States and the Soviet Union of ABM systems (including, under Article 2(c), ABM radars) with the exception of one system to defend the capital city and one system to protect its ICBM silos. A protocol to this treaty, signed in 1974, amended this provision to allow only one system on each side – the United States choosing to protect its missile sites and the Soviet Union choosing to defend Moscow. In the event the United States dismantled its own ABM system in the 1970s while the Soviet Union has maintained, and indeed improved, its ballistic missile defences around Moscow.

Beyond this, the Treaty specified, in Article 5(1) that each signatory undertook 'not to develop, test, or deploy ABM systems which are sea-based, space-based or mobile land-based'. The important omission here, of course, is research. Those who drafted and negotiated the 1972 Treaty recognized that an agreement to abandon research into ballistic missile defence would be impossible to verify. The Treaty, under Article 12(1), provided for verification of compliance by 'national technical means' – in other words, there was to be no international or 'on-site' inspection. It would therefore clearly be impossible, even if it were desirable, to arrive at a verifiable agreement on research.

The inescapable conclusion is that the Strategic Defence Initiative, so long as it is restricted to research, is not in contravention of the ABM Treaty. In any case, United States officials have repeatedly protested that the Soviet Union has already contravened the Treaty by the deployment of mobile ABM systems in the Moscow complex, and more specifically by the construction of a large phased-array radar system at Krasnoyarsk. Article 6(b) of the Treaty binds both parties 'not to deploy in the future radars for early warning of strategic ballistic missile attack, except at locations along the periphery of its national territory and oriented outwards'. The Soviet radar at Krasnoyarsk is certainly not 'along the periphery' of Soviet national territory and, although the Soviet authorities claim that it is an advanced satellite tracking station, its real purpose is not likely to become clear until it becomes operational in 1987 which, significantly, is the date of the next five-year review of the ABM Treaty.

It has also been suggested that the Strategic Defence Initiative might contravene the Outer Space Treaty of 1967 (Appendix 3). This proposition is even more difficult to sustain. Article 4 of this treaty, of which both the United States and the Soviet Union are signatories, forbids states to 'place in orbit around the earth any objects carrying nuclear weapons or any other kinds of weapons of mass destruction, [or to] install such weapons on celestial bodies, or station such weapons in outer space in any other manner'. Clearly research into ballistic missile defence poses no threat to this treaty; nor, indeed, would development and deployment of space-based systems necessarily do so, since most of the technologies at present being considered for strategic defence do not involve nuclear weapons or, indeed, any other weapons of mass destruction.

The only other international agreement occasionally mentioned as a possible bar to the Strategic Defence Initiative is the 1968 Treaty on the Non-Proliferation of

Nuclear Weapons. Indeed, Dr David Owen, the leader of the British Social Democratic Party, suggested in a letter to the Prime Minister, Mrs Thatcher, in 1984, that SDI would be in breach of Article 6 of this Treaty. This article, in full, reads as follows:

> Each of the parties to the Treaty undertakes to pursue negotiations in good faith on effective measures relating to cessation of the nuclear arms race at an early date and to nuclear disarmament, and on a treaty on general and complete disarmament under strict and effective international control.

It is difficult to see what immediate relevance this has to a programme of research into a system designed to provide a defence *against* nuclear weapons, by means of *non-nuclear* technology.

It seems clear, therefore, that until SDI reaches the end of its research phase, and a move into the development and deployment stages is contemplated, no existing international agreement is contravened. Attempts to regulate the military uses of space beyond the scope of existing treaties have not, so far, been attended by any notable success. In 1978 President Carter opened talks with the Soviet Union aimed at preventing the development of space-based anti-satellite systems, and there were bilateral discussions in February and April 1979. Several obstacles quickly emerged. The United States wanted to protect from interference all satellites in which either side 'had an interest' – for practical purposes all existing satellites. The Soviet Union, on the other hand, wanted to restrict this immunity to satellites owned by the United States and the Soviet Union, thus leaving those of China and the NATO allies vulnerable to attack. Furthermore, the Soviet Union wanted to exclude from immunity any satellites performing 'hostile and pernicious acts' that would infringe national sovereignty – a provision open to a very wide spectrum of interpretation, possibly even affecting the activities of reconnaissance satellites. In the event these arguments turned out to be of little more than academic interest, since in 1979 the Soviet invasion of Afghanistan put an end, temporarily, to any serious negotiations.

In August 1981 the Soviet Union took the initiative when its Foreign Minister addressed to the Secretary General of the United Nations a letter containing the following statement: 'The Soviet Union believes that outer space should always remain unsullied and free from any weapons and should not become a new arena for an arms race, or a source of strained relations between states.' Accompanying this apparently unequivocal statement was a draft treaty for consideration by the General Assembly at its 1981 session. At the debate which followed it became clear that although most countries recognized the need to prevent an arms race in outer space, some believed that the draft treaty would not have achieved the aim. The General Assembly, however, adopted a Resolution on *The Prevention of an Arms Race in Outer Space*. It reflected a belief by several member states that the complete demilitarization of outer space, as a short-term goal, was unrealistic, and the UN Committee on Disarmament was requested to consider, at its 1982 session, the possibility of concluding, as the first priority, a treaty specifically directed to the prohibition of anti-satellite systems. In the event, the controversy over intermediate-range nuclear forces in Europe precluded any serious concentration on the problem of weapons in space.

The Soviet Union subsequently submitted another draft treaty to the United Nations, this time calling for the testing and deployment of weapons in space to be banned and all existing anti-satellite systems to be abandoned. This was considered by the First Committee of the United Nations in its 1984 session, but the American view of the Soviet draft was that such a treaty would be difficult to verify and that it was, in any case, not easy to define a 'weapon' in this context. A more basic American objection was almost certainly that such a treaty would prevent the United States from developing an anti-satellite (ASAT) system which would be superior to the existing Soviet system. Some cynical Western observers went so far as to suggest that this was precisely why the Russians submitted the draft in the first place.

Meanwhile, in late 1983, the Soviet Union had broken off all bilateral arms control negotiations with the United States in protest against the deployment in Western Europe of cruise and Pershing II missiles. The Russians stated that they would not resume negotiations until these missiles were withdrawn. In spite of this crude piece of international blackmail, the majority of the Western allies held firm to the deployment programme. In the event, the Russians did return to the negotiating table, albeit with exceptionally bad grace, and on the understanding that, although the substance of the negotiations in Geneva would be the same for the most part, these would be 'new' negotiations – a face-saving formula. The Soviet Union announced that its position now was that there could be no agreement unless the United States abandoned the SDI programme. This technique of demanding substantive concessions as a precondition for entering into negotiations is a familiar one, and it would be very unwise to accept it. Clearly, the Soviet Union is gambling that it can use the SDI programme as a means of dividing the Alliance and damaging the credibility and integrity of the United States – perhaps by offering an attractive set of arms reduction measures while demanding return concessions which it knows the United States cannot grant. For this reason, it has insisted that any progress at the Geneva talks, which began on 12 March 1985, must be paralleled in all three areas of discussion.

It has also sought to strengthen the programme's opponents by threatening to upgrade its offensive forces in order to maximize any elements of vulnerability which might remain in a future American system. Such threats are designed to confirm the hypothesis that the SDI represents no more than a future round of the 'arms race'. As James C. Fletcher, director of the Defensive Technology Study, has testified, 'The ultimate utility, effectiveness, cost, complexity, and degree of technical risk in this system will depend not only on the technology itself, but also on the extent to which the Soviet Union either agrees to mutual defensive arrangements or offensive limitations.'[24]

Yet these are empty threats, intended to disguise the Soviet Union's own work in the field of strategic defence and the already unjustifiable build-up of its offensive forces. The Soviet Union knows that, sooner or later, it must come to terms with the new situation. To do otherwise would be both wasteful and ineffective; to jeopardize the Geneva talks by imposing impossible conditions would be to make the same mistakes as it did in 1983, when Moscow merely talked itself into a corner and damaged its own credibility in the process.

In March 1985, however, negotiations were resumed in Geneva, and many Western observers concluded that the change in Soviet policy had been brought about by President Reagan's Strategic Defence Initiative. Indeed, the Russians had earlier proposed a round of talks designed to concentrate on space-based defence. The United States refused, pointing out that the Soviet Union's own build-up in offensive weapons and ground-based defences had altered the correlation of forces, and that therefore any negotiations on space must be subsumed in general negotiations.

In January 1985 the Russians agreed to this, and the Geneva meetings convened on an agenda which included not only the question of strategic defence, but also those of strategic arms reduction (START) and intermediate-range nuclear forces (INF), dealing with nuclear missiles in the European context. Although the United States has agreed to discuss the question of space-based defences, President Reagan has made clear that SDI is not a bargaining chip to be conceded in exchange for progress on other issues. Indeed Mr Paul Nitze, the US Administration's senior arms control adviser, has said that there would, in any case, be very little room for bargaining on the issue. 'SDI as a research programme cannot logically be limited by agreement,' he said, 'because there is no way you can identify or verify it.' Meanwhile the Soviet Foreign Minister, Andrei Gromyko, has said unequivocally that there can be no progress in Geneva unless the United States renounces its space defence plan.

Whatever may be the impact of the Strategic Defence Initiative on current arms control negotiations, it is clear that the concept on which the initiative is based implies a totally new approach to arms control in the future. If the research programme should prove that a multi-layered defence system is feasible, using the technologies now under consideration, the United States will then have to decide whether to develop, test and deploy such systems. The dividing line between research and development is not, in general, easy to define. When Ambassador Gerard Smith, the chief American negotiator of the ABM Treaty, testified before the Senate Armed Services Committee in 1972 he interpreted the Treaty in these terms:

> The obligation not to develop such systems, devices or warheads would be applicable only to the stage of development which follows laboratory development and testing. The prohibitions on development contained in the ABM Treaty would start at that part of the development process where field testing is initiated on either a prototype or a breadboard [rough working] model. It was understood by both sides that the prohibition on 'development' applies to activities involved after a component moves from the laboratory development and testing stage, to the field testing stage wherever performed.

In other words, there is no restriction on laboratory research of any kind, although this interpretation would evidently inhibit prototype and systems testing as well as engineering development. The United States is in any case already committed to consult its allies before taking this step. Furthermore, such development or deployment would, at present, have to be the subject of consultation with the Soviet Union under the terms of the Agreed Interpretation of the ABM Treaty, by the terms of which both sides agreed to discuss limitations on any ABM systems based on new technologies. If, in spite of all this, the US goes ahead with development and deployment the implications for a wide range of existing arms control will be substantial. In the first place, it would clearly be inconsistent not only with the ABM

Treaty itself, but with the Agreed Interpretation. There is, however, provision in the Treaty for either side to withdraw, giving six months' notice 'if it decides that extraordinary events related to the subject matter of this Treaty have jeopardized its supreme interests'. It would not be too difficult to conceive of a situation in which 'extraordinary events' might be considered by either side to have taken place.

Indeed, as far back as 1972, immediately after the signature of the Treaty, the United States declared unilaterally that 'if an agreement providing for more comprehensive strategic offensive arms limitation were not achieved within five years, US supreme interest would be jeopardized.' It is indeed possible to speculate that the ABM Treaty already has no very great expectation of life.

Development and testing would, however, have a serious effect on other international agreements. The Nuclear Test Ban Treaty of 1963 forbids the testing of nuclear weapons 'in the atmosphere, [or] beyond its limits, including outer space. . . .' As the activation of certain space-based beam weapons, including X-ray lasers, requires an actual nuclear explosion to provide the power source, it is unlikely that such weapons could be developed without tests in space. Furthermore, if such weapons were deployed in space, even without testing, they would contravene the Outer Space Treaty. It seems clear, therefore, that the strategic concept behind the Strategic Defence Initiative calls for an entirely new approach to the doctrines underlying arms control policies.

It is important at the outset to establish a realistic point of departure. Those who refer to the Strategic Defence Initiative as 'the militarization of outer space' seem to ignore the fact that space is already heavily militarized. In addition to ballistic missile early warning systems and satellites for reconnaissance, surveillance and military communications, there are meteorological and navigational systems designed to provide the information necessary for the accurate targeting of strategic offensive weapons systems. Of the 8,000 pieces of hardware that have been sent into orbit over the last twenty-five years, by far the majority were designed for military purposes. There are, at present, more than 1,200 satellites in orbit – 800 from the Soviet Union and 400 from the West – most of them with a military application. Space is already the 'high ground' of contemporary strategy. Future arms control negotiations which do not take that fact into account are unlikely to be productive.

In considering the impact of the SDI on arms control, it is essential to dismiss the fallacy that all arms control agreements are necessarily desirable. As Seymour Weiss, former Director of Politico-Military Affairs in the State Department has pointed out, if an arms control agreement is clearly either unverifiable or unenforceable, it serves Soviet purposes. Arms control should be a facet of defence policy, and agreements that do not enhance security are worse than no agreements at all.

9

Strategic Defence and NATO

Of all the anxieties aroused by President Reagan's March 1983 speech, those of the European members of NATO were possibly the most acute. The vision of the United States (and eventually the Soviet Union) protected by an invulnerable defence against ballistic missiles, while Europe remained open to nuclear devastation, revived all the fears, never far below the surface, of 'decoupling' and American isolationism.[25]

Concerns of this kind have been increasing in recent years. The growing concentration of United States foreign policy on Latin America; the doubts cast by Dr Henry Kissinger and others on the continuing validity of 'extended deterrence'; and the frustration in Washington at the apparent failure of the Western Europeans to take their own defence problems seriously; all these factors have combined to create among the European allies an apprehension that the United States might be irresistibly attracted by a policy of global unilateralism, in which the security of Western Europe would become only one of a number of American concerns – not necessarily high on Washington's list of priorities.

Among the Western Allies, two had special concerns about the Strategic Defence Initiative: the United Kingdom and France. In a situation in which both superpowers had developed an effective defence against ballistic missiles, the nuclear strike forces of both countries would clearly lose some, if not all, of their credibility as deterrents – especially against conventional attack by the Soviet Union.[26] It was not, therefore, a matter for great surprise when the West European Allies reacted in a characteristically West European fashion. Each country instinctively consulted its own interest and exhibited the same signs of confused agitation that they have been accustomed to show when confronted with any new political or military development.

The French Defence Minister, Charles Hernu, expressed dark misgivings about 'the militarization of outer space' at a meeting of NATO defence experts in Munich, during which the French President, François Mitterrand, spoke of the possibility of Western Europe setting up its own space surveillance system. Foreign Minister Roland Dumas, referring to fears about the continuing credibility of the French and British deterrents, said that these two countries were 'not worried, for they are convinced that they have fifteen or twenty years ahead of them to reflect'. As the debate progressed the French attitude towards SDI, at first deeply suspicious, seemed to grow more positive.

Meanwhile in Bonn a predictably cautious line emerged. Chancellor Kohl, at the defence meeting in Munich, expressed lukewarm support for the SDI; President von Weizsäcker, in a somewhat gnomic speech at the Königswinter conference in March 1985, appeared, so far as his words could be clearly interpreted, to be expressing grave reservations about space-based defence. Herr Genscher, the German Foreign Minister, was also unenthusiastic, while Manfred Wörner, the Defence Minister, expressed some fundamental criticisms of the plan. In general the German view seemed to be based on the assumption that President Reagan would go ahead with the research programme, whatever the opposition, and that Germany should not be seen to stand out as an opponent.

Signor Bettino Craxi, the Italian Prime Minister, assured President Reagan in the course of a visit to Washington of his 'full understanding' of the Strategic Defence Initiative. The British Prime Minister did the same, while extracting from the American President a four-point agreement that the US and Western aim was not to advance superiority, but to maintain balance; that deployment of strategic defence systems would be a matter for negotiation; that the overall aim was to enhance, not undercut deterrence; and that

East–West negotiations should aim to achieve security with reduced levels of offensive systems on both sides. These undertakings, which were no more than a restatement of the known position of the United States Administration, enabled Mrs Thatcher to announce that Britain was 'at one' with the United States on SDI. The *Gemütlichkeit* was somewhat disturbed when Sir Geoffrey Howe, the Foreign Secretary, delivered his speech at the Royal United Services Institute on the eve of the opening of the Geneva arms control negotiations (see page 18).[27] Although the speech itself was no more than a catalogue of the questions which most experts had been posing for some months on the subject of the SDI, it succeeded in attracting for Sir Geoffrey the anger of American officials, a magisterial reprimand from *The Times*, and the gleeful approval of the Kremlin.

Meanwhile, outside Europe, the West, in its broader politico-strategic sense, was equally confused. Mr Yasuhiro Nakasone, the Japanese Prime Minister, expressed his 'understanding' of the SDI, but doubts were expressed in Tokyo regarding the implications for Japan of a programme which might include research and development of nuclear-powered lasers. In the Canadian House of Commons Mr Joe Clark, Secretary of State for External Affairs, said that 'in the light of significant Soviet advances in ballistic missile defense research in recent years, and deployment of an actual defense system, it is only prudent that the West keep abreast of the feasibility of such projects', although he went on to issue the standard caveat that deployment would be in breach of the ABM Treaty.

It was left to Lord Carrington, the Secretary General of NATO, to offer a few characteristically commonsense remarks on the whole issue. Based on his own understanding of the theory of multi-layered defence, his conclusion was that the United States was 'absolutely right' to embark on the research programme, and pointed out that if the Americans failed to do so and 'we suddenly woke up and found that the Soviet Union could do what the Americans are seeking to do', nobody would be more critical than those Europeans who are now criticizing the Americans for engaging in the research. In an impromptu and typically pragmatic comment, which may lack something in strategic sophistication, he commented: 'In any event, I think it is quite a good idea to shoot down missiles with nuclear warheads wherever they are going. It would certainly be to somebody's benefit if you succeeded in doing it, whoever it happens to be.'

This breezily expressed implication that the SDI might conceivably benefit Western Europe as well as the United States may have helped to concentrate the minds of some of the sceptics in Western Europe. That was certainly the effect of an announcement by Mr Caspar Weinberger at a meeting of the NATO Nuclear Planning Group in Luxemburg on 26 March 1985, when he invited the Western Allies to participate in the programme of research. In a letter to each of the NATO defence ministers, he invited them to identify areas of technology in which their industries could best contribute to the research programme. Similar invitations were sent to France, Japan, Australia and Israel. Only Australia declined out of hand. There is, it seems, nothing like the prospect of a few lucrative high-technology contracts for overcoming political sensitivities and ideological doubts.

Despite these obvious benefits, the substantial efforts made by the Reagan administration in the early part of 1985 to put its case across to the NATO governments reflected a growing concern that the SDI programme would be used to force a division between the United States and its European allies. There were heated exchanges between the US Assistant Secretary for Defense, Richard Perle, and the British Foreign Secretary, Sir Geoffrey Howe, after Sir Geoffrey had expressed reservations about the programme. The Foreign Secretary's speech to the Royal United Services Institute has not done as much damage to British–American relations as had earlier been feared. In many quarters its timing was regarded as a deliberate attempt by the British Foreign and Commonwealth Office to weaken the American position at the current arms negotiations in Geneva. It was also seen as a reflection of the somewhat quixotic ambition of the British Government to exercise a mediating role in superpower relations.

West European leaders have more recently shown themselves willing to support the SDI research programme on the basis of the 'four points' agreed by Margaret Thatcher and President Reagan at Camp David in December 1984. Other tentative supporters have emphasized the need for stability to be maintained during any transitional period, for the Alliance to be consulted at each stage, and for the implications of the SDI for deterrence and conventional defence to be studied with careful deliberation.

The West German government, for its part, has pledged support for the SDI 'as long as no other bilateral agreements have been reached' and that participation in the programme enables its country to keep abreast of technological progress. Italy has also affirmed its support, on condition that 'defence and peace, based on a balance of forces' are assured. The NATO Defence Ministers, meeting in Luxemburg in March 1985, also welcomed the research programme as being 'in NATO's interests'. Lord Carrington has urged a consensus on the continuation of research, with the provision that there will be a 'firebreak' between this and any decision to deploy, and that 'the present strategy of nuclear deterrence will be maintained in the interim'.

Britain and France, however, are known still to be worried about the implications of the SDI for their own independent nuclear deterrents – even though an effective strategic defence on both sides is many years away and it will always be necessary to preserve an element of nuclear deterrence in any comprehensive system. The French President, François Mitterrand, and other prominent Europeans have expressed deep concern, much of it conditioned by poor official advice, the doctrinal bias of government and expert opinion, and an ignorance of technological developments. Clearly, the possibility of serious friction within the Western Alliance will exist for some time. In these troubled waters the Soviet Union will be fishing assiduously.

Behind the suspicion and confusion among the European

Allies there are some genuine anxieties about the implications of the Strategic Defence Initiative. The principal concern is based upon the belief that the technology for defence against intercontinental ballistic missiles will not be effective against shorter-range missiles such as the SS-20. Yet this is by no means certain. Dr George Keyworth, the Science Advisor to the US President, has said that it may well be possible to intercept SS-20s in the early phases of their flight, although he notes that certain elements of any system designed to defend Europe against them would have to be based in Europe itself. The inference from this is that technologies which prove to be effective for the defence of the United States would also be effective for the defence of Western Europe, provided that the European Allies were prepared to accept the political and economic implications of stationing part of the defensive system on European soil.

However, doubts have also been expressed about the value to Europe of a system which is not designed to intercept nuclear delivery vehicles such as aircraft, artillery rockets or cruise missiles, and which will therefore not eliminate a substantial part of the nuclear threat to Western Europe. It is true that the multi-layered defence system now envisaged would be ineffective against air-breathing delivery systems such as bombers and cruise missiles, and also against artillery shells and probably against submarine-launched missiles fired from very short range. To exaggerate the importance of this, however, is to ignore the fact that those are not first-strike weapons – and it is primarily to deal with a first strike that the Strategic Defence Initiative has been conceived. Furthermore, an effective defence against bombers and cruise missiles is by no means out of the question in the same timescale as the SDI.

The principal strategic objection advanced by Europeans to SDI is that it would, if effective, remove the threat of escalation and thus destroy the concept of flexible response. This seems a somewhat perverse argument. If flexible response, with its essential corollary of an American nuclear guarantee, ever had any real credibility, that credibility would surely be enhanced by reducing the vulnerability of the American retaliatory force. Furthermore the major targets of Soviet intermediate-range missiles such as the SS-20 are NATO's conventional infrastructure – depots, ports, airfields, command centres, and communications; if these could be defended, even partially, by an air defence system reinforced by a missile defence system, the conventional capability of NATO would be significantly enhanced.

In an address to the American Society of Newspaper Editors in April 1985, Caspar Weinberger, the US Secretary for Defense, underlined the American view of this problem:

> Most important, our decision to embark on this research effort does not mark an abandonment of our commitment to the security of our allies. Effective defense against ballistic missiles would enhance considerably the security of our friends and allies by protecting them as it would protect us from the threat of Soviet missiles. We define an effective defense as one that would destroy, by non-nuclear means, intermediate as well as intercontinental Soviet missiles. When President Reagan announced the Strategic Defense Initiative two years ago, he reiterated – as he has so often – the inseparability of our allies' security with that of the United States. He said: 'Their safety and ours are one, and no change in technology can, or will, alter that reality.'
>
> In enhancing NATO's ability to deter nuclear attack, such strategic defenses could also strengthen our ability to deter conventional Warsaw Pact aggression. SDI could reduce the ability of Soviet ballistic missiles to place at risk those facilities essential to a conventional defense of Europe, facilities such as airfields, ports, depots, and communications facilities. An effective defense against ballistic missiles would create great uncertainties for the Soviet military planner. It thereby would reduce the likelihood the Soviet Union would contemplate any kind of attack in the first place.

Most European objections to the Strategic Defence Initiative are based on one or two possible contingencies. In one, the United States has an effective ballistic missile defence system, but the Soviet Union has not. This gives rise to fears of instability in the strategic balance and the withdrawal of the United States into 'Fortress America'. In the other both the United States and the Soviet Union have effective strategic defences. This gives rise to fears of a superpower 'stand-off' and the possibility of a limited nuclear war in Europe. There is, however, a third scenario to which Europeans tend to devote far less attention. It is one in which the United States is persuaded, for reasons of cost or political expediency, to abandon research into strategic defence, while the Soviet Union goes on to develop an effective system.

The implications of this should provide far more cause for concern than the other two possibilities. The Soviet Union has already built up an offensive nuclear missile force with a much greater first-strike potential than that of the United States. If it were to add to that a unilateral capacity to defend itself effectively against a large proportion of the American missiles force, the possible consequences for the West would be, to say the least, grave. If the conduct of Soviet foreign policy since World War II is any guide, there is a high risk that the Soviet Union would use such a first-strike capability to exert irresistible political pressures on Western Europe. Finlandization would be an established fact.

It is, however, extremely unlikely, whatever may be the pressures, within the United States, from Western Europe or from the Soviet Union, that President Reagan will abandon the research programme; and this confronts the European Allies with two major decisions. The first is whether, and how, to participate in the Strategic Defence Initiative; and the second is how to remedy the imbalance in conventional forces, which will become a crucial issue whether the United States alone, or the United States and the Soviet Union together, develop an effective system of ballistic missile defence.

10
The Defence of Western Europe

In the course of his March 1983 speech President Reagan made a significant statement which, in the instant Star Wars uproar, largely escaped attention. After proposing a programme of research into ballistic missile defence he went on to say:

> At the same time we must take steps to reduce the risk of a conventional military conflict escalating to nuclear war, by improving our non-nuclear capabilities. America does possess – now – the technologies to attain very significant improvements in the effectiveness of our conventional non-nuclear forces. Proceeding boldly with these new technologies we can significantly reduce any incentive that the Soviet Union may have to threaten attack against the United States or its allies.[28]

The so-called Star Wars speech was, in fact, something much more than a new approach to nuclear deterrence; it was a signpost to a completely new approach to the strategic doctrines upon which Western defence and security are based. In addition to identifying the moral and prudential weakness of a deterrent posture based solely on a suicidal threat of retaliation, President Reagan at the same time underlined the weakness of a conventional defence doctrine based on forces inadequate to carry it into effect.

The President was, in fact, articulating a truth about the defence of Western Europe which has for some time been a matter of concern to those most directly concerned in it. A certain prescription for disaster is to base national or collective security upon a carefully constructed military strategy, and then persistently fail to provide the resources necessary to implement it. It is for this reason that the ability of NATO to implement the defensive-deterrent doctrine generally described as 'flexible', or 'graduated response', has for some time been regarded with considerable reservation by a substantial body of opinion – not only among professional planners and military commanders but also among academic strategists.

Perhaps the most articulate and authoritative advocate of more realistic approaches to the defence of Western Europe is General Bernard Rogers, Supreme Allied Commander Europe, who has raised to an almost evangelical level his campaign to exploit the Western technological superiority referred to in the President's speech in order to offset the numerical superiority of the Warsaw Pact, and so improve the deterrent posture of NATO's defensive forces.[29] He proceeds from an assumption that the doctrine of flexible response is entirely valid, provided that the resources necessary to implement it are clearly available.[30] The validity of this doctrine depends upon the concept of controlled escalation, in which any attack by the Warsaw Pact forces would be met at each stage by an appropriately graduated reaction, including the use of 'battlefield' or 'tactical' nuclear weapons if it proved impossible to contain the enemy advance by conventional means. This has involved the forward deployment of a considerable number of low-yield nuclear weapons to implement the first stages of the process of escalation, and, by extension, to provide a manifest deterrent against an attack with even limited objectives.

This doctrine has always, of course, implied a readiness by NATO forces to use nuclear weapons first; and it is in this context that most of the doubts about its credibility have arisen. Many serious analysts have insisted that, as tactical or battlefield nuclear weapons are for all practical purposes under ultimate American control, their first use against a conventional attack would be extremely unlikely, since there would be no guarantee of a similarly limited Soviet response. In the language of the more trenchant critics, no President of the United States is likely to put New York or Chicago at risk to preserve the integrity of the European theatre of operations. The counterargument has been that no leader of the Soviet Union could ever be sure of this, and that the inevitable doubts about American reaction were in themselves an effective deterrent. As General Rogers insists, even with adequate conventional capabilities, NATO could never be certain of defeating a conventional attack without escalation; removing the element of uncertainty from the mind of a potential aggressor by declaring a 'no first use' policy would therefore seriously weaken the nuclear deterrent.

The Warsaw Pact, however, has always been able to outrange NATO's 'battlefield' nuclear weapons with its conventional artillery. As Dr Manfred Wörner, Defence Minister of the Federal Republic of Germany, pointed out in May 1982, 60 per cent of the United States nuclear weapons in Europe have a range of less than 19 miles and the great majority of these weapons have a range of under 9 miles.[31]

> This means nothing more and nothing less than that the greater range of Soviet tube and rocket artillery presents the opportunity for the Warsaw Pact, under many battle conditions, to destroy by conventional means NATO's nuclear insertion capability, while its own weapons are beyond the effective range of NATO forces.

The equation has been further complicated by significant improvements in the nuclear capability of the Soviet Union. The strategy of flexible response was developed over a period of fifteen years during which the United States possessed a clear degree of nuclear superiority. It was estimated in 1952 that the conventional defence of Western Europe required force levels of ninety-six divisions and 9,000 aircraft to pose a credible response to the powerful ground forces of the Warsaw Pact. In the political and economic climate of post-war Europe these levels (the so-called 'Lisbon Goals') were clearly unrealistic, and even a compromise plan for fifty divisions and 4,000 aircraft had no prospect of being achieved.

By 1956 NATO had decided to settle for twenty-six divisions and 1,400 aircraft.[32] These demonstrably inadequate conventional forces were to act as a 'tripwire', designed to trigger off United States nuclear retaliation against nuclear attack. It was planned to deploy 15,000 'tactical' nuclear weapons which were to be, in effect, an extension of conventional firepower. It was still the *strategic* superiority of the United States which was to provide a nuclear umbrella of extended deterrence for Western Europe. In the event only 7,000 battlefield nuclear weapons were deployed, and in 1967 NATO adopted its strategy of flexible response, which postulated a significant improvement in conventional forces – an improvement which never materialized. Theory and fact therefore failed to coincide. There was, in fact, no capacity for flexible response: the security of Western Europe continued to rely upon a form of massive retaliation, which, so far as it ever made sense at all, did so only in the context of decisive American nuclear superiority.

In the meantime, that superiority has virtually disappeared. As Sam Nunn, the United States Senator from Georgia, pointed out in 1982 in the course of his testimony to the Senate Armed Services Committee, 'By attaining strategic nuclear parity with the United States, the Soviet Union has severely undermined the credibility of US strategic nuclear forces as a deterrent to a conventional attack on Europe.'[33]

Many serious observers believe that Senator Nunn was understating the case, and that in some respects the Soviet Union now enjoys a measure of nuclear superiority over the United States. In addition its conventional predominance has increased in quantitative terms; while Nato's ability to compensate for numerical inferiority with qualitative superiority has been substantially eroded. Finally, the Soviet Union has also achieved a clear superiority in theatre nuclear forces with the development and deployment of the SS-20, 21, 22 and 23 as well as a force of nuclear-capable attack aircraft, some of which (for example, the Backfire) are superior to anything NATO possesses in term of range and payload.

These recent developments have been instrumental in

creating severe political problems within the Western Alliance. West European leaders have begun to fear that the ability of the Soviet Union to inflict, with such weapons as the SS-20, massive damage on targets in Western Europe without threatening the continental United States, might lead to a 'decoupling' of the American nuclear guarantee. In this contingency, there would be a real danger of Western capitulation to Soviet military superiority. The classic scenario of defeat 'without a shot being fired' would then become a serious possibility. This fear has been aggravated by public speculation about another scenario, namely the possibility of a 'limited' nuclear war in Europe which would leave the United States intact. The first reaction to this somewhat improbable contingency took the form of a demand from Western European political leaders that the United States should counter the threat of the new generation of Russian theatre nuclear weapons with a programme of 'theatre force modernization'. This eventually emerged in the form of the decision to station American ground-launched cruise missiles and Pershing II ballistic missiles on the soil of certain European members of NATO.

The implementation of this programme proved to be the catalyst for a massive resurgence of anti-nuclear protest in Western Europe. Describing themselves somewhat obscurely as peace movements, a coalition of misguided idealists, with a sprinkling of useful idiots and Soviet agents (conscious and unconscious) began to conduct campaigns, often involving violence and civil disobedience, designed to bring about unilateral nuclear disarmament by the Western Allies, cancellation of the theatre force modernization programme, and the closing down of American bases in Western Europe. In the margin of the manifestations there have been a number of proposals, superficially more reasonable but in effect almost equally dangerous, for concepts such as nuclear freezes, 'no first use' declarations, European nuclear-free zones and postponement of new theatre-based systems pending a resolution of East–West negotiations on intermediate-range nuclear forces.

It is therefore not too extreme to suggest that a strategy of flexible response, if it is widely perceived to be incapable of implementation, carries with it dangers that might eventually threaten the disintegration of the Western Alliance. So long as NATO relies on defensive resources which allow virtually *no* flexibility and no range of options between nuclear retaliation and retreat (between 'suicide and surrender') there is a very real danger that, in a crisis, the West might be forced to submit to military blackmail; and the growing public awareness of these dangers is beginning to cause severe political problems in several NATO countries.

Yet the obvious alternatives present their own dangers and difficulties. To return to the declared policy of massive retaliation, in the current state of the nuclear balance, would obviously lack all credibility; while to attempt to match Russian conventional strength by building up NATO's own conventional forces to an adequate level is as unrealistic, politically and economically, as it was at the time of the Lisbon Goals. A new conceptual approach to the defence of Western Europe is clearly essential; and in this context modern technology offers remarkable possibilities for the strengthening of conventional striking power without substantial numerical increases in forces and equipment.

The first priority is to acquire for NATO forces the flexibility and room for manoeuvre which is conspicuously absent from existing defensive plans. In terms of time and space, this must involve an extension of the battlefield which will deny to the enemy the freedom to conduct the land-air battle according to his preferred tactical doctrine and on ground of his own choice. NATO's current defensive plan is based upon the assumption that the initial conventional battle will be fought on West German territory, principally against the assault echelons of the Warsaw Pact forces. To provide the necessary room for manoeuvre on the soil of the Federal Republic clearly would carry with it political implications of a quite unacceptable kind – as well as a number of manifest military disadvantages. The current NATO defensive plan is therefore based on a 'forward strategy', theoretically designed to meet and destroy advancing Warsaw Pact forces in a comparatively narrow zone west of the River Elbe.

As a result the tactical planners of the Soviet Union have evolved a system of carefully structured echelons, designed to provide continuous reinforcement at the chosen points of attack. This method is fully documented in the West: the most succinct unclassified description is contained in a document, *Soviet Army Echelonment and Objectives*, prepared for a conference of Commanders of the Central Army Group of the Allied Forces in Europe held in October 1978.[34] The echelon system, which is applied at all organizational levels from front down to regiment, is not to be confused with the system of reserves familiar in Western armies. The Soviet second echelon is an integral part of a combat formation not directly participating in the battle at any given moment. It is usually about one-third of the size of the first echelon, and its principal purpose is to intensify pressure on the main axis of advance and to exploit success in depth. Reserves, in the Western sense, are held *in addition to* the second echelon.

In the context of existing NATO defensive plans, the Warsaw Pact would be able to implement this doctrine without significant hindrance from Western conventional forces. The need to repair this deficiency has already been the subject of extensive analysis, especially in the United States; and there has been much debate within NATO concerning various proposals for the improvement and upgrading of conventional defence in Europe. These ideas have found political expression in the statements of Senator Nunn and Dr Manfred Wörner; and they have received public support from General Rogers. Furthermore, they have been reflected in *Airland Battle 2000*, a recent attempt by the United States Army to formulate new conventional defence concepts on the assumption that 'a linear defense doctrine which is designed to prevent breakthrough by means of battles of attrition against a vastly superior force is militarily unsound.' *Airland Battle 2000* does not, however, totally satisfy the preoccupations of the Supreme Allied Commander Europe, who sees it as a potentially useful national concept which needs to be harmonized with the basic NATO doctrine of attacking Warsaw Pact follow-on forces using such devices as 'smart' terminally guided sub-munitions (see page 125). As General Rogers insists, the theory of the extended battlefield needs to be considered in the general context of NATO's official strategy of flexible response.

The concept of the extended battlefield may be briefly described as the expansion of the battlefield, in the event of a Warsaw Pact attack, deep into the territory of the attacker. This would involve the destruction of enemy air forces on the ground; the interdiction of second-echelon reinforcement by conventional attack on fixed targets such as bridges, assembly areas, ammunition dumps and fuel depots; and effective conventional attacks on moving formations anywhere in the battle area. In order to implement this concept it will be necessary to apply a wide range of high-technology weapons systems, as the American General Starry has pointed out in an article in *Military Review* of March 1981, 'not just weapons of greater lethality and greater range, but automated systems and communications systems for more responsive command control, as well as sensor systems to find, identify, and target the enemy and to assess the effectiveness of deep attack.'[35]

As far as the first two elements in the extended battlefield are concerned – destruction of enemy air forces and interdiction by destruction of fixed installations – the targeting problem is relatively simple, since the location of principal targets is already known. The main counter-air targets are the Main Operating Bases from which Warsaw Pact offensive operations against NATO's rear areas would be conducted; and the main interdiction targets are bridges and other 'choke points' through which reinforcements and logistic support would have to pass. The problem, therefore, is not one of finding these targets but of destroying them *within the first day of operations*. NATO's present tactical air forces are inadequate to fulfil this task – many tactical aircraft are without night and all-weather equipment – and in any case close support and other responsibilities would not leave sufficient resources for effective sorties against fixed installations.

The obvious instrument for the destruction of fixed targets is the unmanned platform: cruise and ballistic missiles equipped with precision-guided munitions, many of which are now accurate to within a few yards. These would be used to attack Main Operating Bases and thus force the returning first wave of attacking aircraft to return to Dispersed Operating Bases; and also to destroy important choke points. Ground-attack aircraft armed with advanced area munitions could then be used to destroy enemy aircraft on the less well defended Dispersed Operating Bases and to attack ground forces held up in the interdicted choke points.

The more difficult and complicated problem arises in the third element of the general concept: the destruction of mobile forces on the battlefield. The essential requirement is one of 'real-time' intelligence to provide an integrated system of surveillance, tracking and targeting. The 'state of the art' in information technology has now brought this well within the reach of battlefield commanders at all levels. The use of microprocessors, radio-emitter location systems and advanced electro-optical imaging techniques can provide a distributed, and survivable, target-acquisition capability down to battalion level. At the same time there are improvements in weapons systems which can dramatically improve the effectiveness of ground-attack aircraft. The first generation of these weapons includes passive systems dispersed in the form of anti-armour mines delivered by stand-off weapons fitted to NATO's existing fighter-bomber aircraft. A generation of terminally guided anti-armour sub-munitions, available in the second half of the 1980s, has been somewhat colourfully, but accurately, described by an American defence journalist:

> Each missile, carried by aircraft or flying on its own would have a 'bus' at its front end that carried up to 25 'sub-munitions' or small missiles of 5 to 6 inches diameter. Cut loose in the target area, the sub-munitions would have sensors 'smart' enough to distinguish

> between a hot spot on a tank and a decoy flare or some other distraction. The target seekers on these little missiles would move their control surfaces and steer them straight to the bull's eye – one shot, one kill.[36]

According to a West German study group, the effect of these systems on the capability of conventional forces would be dramatic. They calculate that to destroy 60 per cent of a Soviet division (and thereby to make it ineffective in battle) would require, with existing weapons, over 2,000 manned aircraft sorties or 10,000 conventional missiles. With unguided stand-off weapons these figures would be reduced to 300 and 1,500 respectively; and using guided submunitions only fifty to sixty missiles would be needed. In other words, using modern weapons technology, conventional systems can enter 'a dimension that heretofore was limited to nuclear weapons'.

It is, however, important to emphasize that this potential for a significant technological upgrading of non-nuclear defence cannot provide for NATO what has been vaguely described as a 'conventional option'. Not only will the strategic nuclear forces of the United States, together with any strategic defence systems which might be deployed in the future, continue to provide the ultimate deterrent against the exercise by the Soviet Union of many of its military options; shorter-range nuclear weapons will also continue to be essential in the NATO inventory, since, with the threat of their use removed, the Warsaw Pact forces would be able to adopt tactical methods involving far greater concentration and economy of force. This carries with it the clear corollary that NATO must retain the option of first use. It is one thing to seek a defensive posture in which the nuclear threshold is significantly higher and in which the possible need to use nuclear weapons first is correspondingly reduced. It is another matter altogether to engage in anything as quixotic as a 'no first use' declaration. As *The Economist* argued (31 July 1982), 'These things sound marvellous when they are first said but they are, after all, only declarations –

uncheckable, unenforceable and, when a healthy realism asserts itself, unbelievable.'[37]

The technological improvement of Western conventional defences would not, on the other hand, be entirely without its impact on the climate of nuclear arms control. It would be possible – indeed desirable – to reduce and possibly eliminate the role of the so-called 'battlefield' nuclear weapons which exist only on the premise of a dangerously low nuclear threshold; and possibly even to reduce (in exchange for substantial Russian concessions) the number of cruise and Pershing II missiles planned for deployment in Western Europe. It is, however, important to bear in mind that, as General Franz-Joseph Schulze, former Commander-in-Chief Allied Forces Central Europe has written,

> The security of Western Europe can only be safeguarded by the close and indivisible coupling of the conventional and nuclear forces in Europe with the strategic nuclear potential of the United States. Intermediate-range nuclear systems on European soil are a visible demonstration of US resolve to confront the Soviet Union with the incalculable risk that her territory would not remain a sanctuary if she initiated a military conflict in Europe.

This is, of course, even more important in the context of the SDI. A possible model for the restructuring of nuclear forces in Europe has been proposed by Donald R. Cotter, a former Assistant to the US Defense Secretary and now Director of the Center for Strategic Concepts – the *NATO Nuclear Overwatch Force*. According to this concept, battlefield nuclear weapons would be replaced by improved conventional systems such as the Multiple-Launch Rocket System with terminally guided weapons, together with Quick-Reaction missiles and air forces. The nuclear element of the deterrent would consist of the 464 ground-launched cruise missiles now being deployed, with possible variations including the basing of some cruise missiles at sea on submarines or surface ships, and in the air, possibly in helicopters.

These are, however, like the Strategic Defence Initiative,

relatively new concepts, and there is in NATO a predictably uncoordinated approach to their implementation. The Alliance now finds itself in an almost unique situation in which military and political imperatives point in the same direction; it also finds itself in a situation, unhappily far from unique, of political paralysis. As Senator Nunn has pointed out, 'the NATO alliance is in need of major repair, militarily, politically and economically'. A valuable first step towards effecting that repair would be to abandon a defensive posture that has been suspect since the United States lost the monopoly of strategic nuclear power; and which has now become, quite simply, politically indefensible and militarily incredible.

11

A New Strategic Concept

President Reagan's Strategic Defence Initiative provides the catalyst for an entirely new approach, in the period between now and the end of the century, to the central political and ideological confrontation: that between the Soviet Union and its allies on the one hand and the United States and its allies on the other.

In an important article published in *Foreign Affairs* in April 1985, Fred Charles Iklé, United States Under Secretary for Defense for Policy since 1981, set out the agenda for this approach:

> We need to accomplish a long-term transformation of our nuclear strategy, the armaments serving it, and our arms control policy. To begin with, we must disenthrall ourselves of the dogma of consensual, mutual vulnerability – the notion that unrelieved vulnerability of the United States and the Soviet Union to each other's nuclear forces is essential for halting the competition in offensive arms, and is the best guarantee against the outbreak of nuclear war.
>
> Such a relationship never obtained in the past and it is most unlikely to come into existence in the future. Moreover, an agreed balance of mutual vulnerability would be repugnant on moral grounds. And it could in no way reduce the consequences of an accidental nuclear attack. In the long run, reliance on a balance of mutual vulnerability would favor totalitarian regimes, with demoralizing effect on democracies.
>
> The key now for the needed transformation is technological development to make effective defensive systems possible for the United States and our allies. The priority requirement is non-nuclear missile defenses capable of negating the military

utility of a Soviet missile attack and of diminishing its destructiveness. Depending on progress in arms reductions, the missile defenses might later be complemented with air defenses. As strategic defenses make it increasingly unlikely that Soviet offensive forces can accomplish their mission, the incentive for new Soviet investment in them is reduced. We thus enhance Soviet willingness to join us in deep reductions of offensive forces.

To this end we ought to take two complementary approaches. We should energetically seek Soviet cooperation, since it would greatly ease and speed the transformation. But we must also be prepared to persist on the harder road, where the Soviet Union would try as long as possible to overcome our defenses, and would resist meaningful reductions in offensive forces. The better prepared we are and the more capable of prevailing on the hard road, the more likely it is that the Soviet Union will join us on the easy road.

It is important to recognize, then, that the confrontation exists. We are engaged with the Soviet Union in a continuing conflict of faith, of ideas and of moral values. It is a conflict between two totally irreconcilable political systems, one in which the individual exists only to serve the state, and one in which the state exists to serve the individual. These two systems might conceivably co-exist in something resembling peace and stability, but there can never be any compromise with the values of totalitarianism. It is entirely fanciful to suppose that in the next ten or fifteen years some miraculous transformation will occur, and that the confrontation between Soviet imperialism and the free world will cease to exist.

As long as the Soviet Union maintains its great apparatus of military power, and demonstrates at frequent intervals the readiness to use it ruthlessly in pursuit of its foreign policy, the free world must make prudent arrangements for its security and survival. In the existing circumstances, the primary security objective of the West must be to reduce to a minimum the risk of war, while preserving our right to live in freedom. In pursuit of these twin aims, we have, since the present confrontation emerged after World War II, based our security policy on deterrence. Deterrence itself is not a new concept; it has always been at the heart of any defence against the use of military force. It is the prevention of conflict by convincing a potential opponent that the risks and costs of aggression would far outweigh any conceivable advantage he might hope to gain.

The introduction of the nuclear weapon into this equation has affected it in two ways. The first is to make the cost of aggression so high and so appalling that the deterrent content of a military establishment which includes nuclear weapons is substantially more potent than one which does not; but, secondly, the implications of a *failure* of deterrence are too terrible to contemplate. In previous wars, when the deterrent power of armies, navies and, more recently, air forces has failed, the resulting conflict has usually been wretchedly damaging and destructive; but it has not been irretrievably annihilating. A nuclear war would be. It is for this reason that the posture of Mutual Assured Destruction is, and always has been, a dangerous and unsatisfactory concept. It has, it is true, prevented a war between the two great conflicting political and economic systems for forty years; but the potential cost of a breakdown is so crushing that it would be irresponsibly foolish not to seek constantly for some more reliable means of ensuring peace with security. As *The Times* declares, '. . . it must be right to prefer a defensive system, albeit an imperfect one, than to continue with the arid increase of mutual assured destruction.'

The suggestion that a strategic defence system would constitute a move away from the policy of deterrence is based on the false belief that the threat of nuclear retaliation is the *only* effective deterrent. A potential aggressor can also be deterred from military action if he is faced with defences strong enough to deny him whatever advantages he might have hoped to gain. The Strategic Defence Initiative, provided that it is regarded as a coherent strategic concept, provides a blueprint for a deterrent based on effective defences rather than on a suicide pact. The tendency has been, especially in the press, to concentrate on space-based defence against nuclear weapons to such an extent that the corollary to it, referring to defence against conventional attack, and which formed an equally important part of the President's speech, has been for all practical purposes entirely ignored.

A brief conceptual framework within which it would be possible to develop a new strategic relationship between the superpowers and their allies was set out by Paul Nitze, Special Advisor on arms control to the President and the Secretary of State in February, 1985. It had been evolved specifically for the meeting in January between the American Secretary of State George Shultz and the Soviet Foreign Minister Andrei Gromyko, and in preparation for the arms control negotiations which began in Geneva in March. Mr Nitze said:

> During the next ten years, the US objective is a radical reduction in the power of existing and planned offensive nuclear arms, as well as the stabilization of the relationship between offensive and defensive nuclear arms, whether on earth or in space. We are even now looking forward to a period of transition to a more stable world, with greatly reduced levels of nuclear arms and an enhanced ability to deter war based upon an increasing contribution of non-nuclear defenses against offensive nuclear arms. The period of transition could lead to the eventual elimination of all nuclear arms, both offensive and defensive. A world free of nuclear arms is an ultimate objective to which we, the Soviet Union, and all other nations can agree.

It has to be said at once that this formulation contains one critical weakness – in its last sentence. It is by no means self-evident that 'all other nations can agree' to a world free of nuclear arms. Not all other nations will be able to provide for themselves the kind of high-technology conventional defences needed to deter potential aggressors. It is, for example, doubtful whether the People's Republic of China would be content to abandon its nuclear strike force while a million Soviet troops stand on its border. Nor does the strategic concept outlined by Mr Nitze take account of terrorist states which might seek to acquire nuclear weapons (possibly of a kind which could not be dealt with by strategic defences) for purposes entirely unconnected with their national security.

These are, however, problems which will have to be dealt with in any case, and they need not prevent a serious consideration of the basic strategy. The new concept does not provide any 'quick fix' or instant solution to the present strategic confrontation. It seems clear that for some years – probably for the rest of the century – deterrence will continue to be based ultimately on the threat of nuclear retaliation. The principal and urgent concern therefore is in the field of arms control. Offensive nuclear systems on both sides exceed the requirements of an effective deterrent, and it is in this aspect of the problem that fears of a 'first strike' begin to arise. There is therefore a clear need for *verifiable* agreements to reduce the existing numbers of strategic and intermediate-range nuclear weapons.

This is easy to propose, but it will be by no means easy to bring about. The Soviet Union has already stated categorically that there can be no agreement of any kind on arms control while the United States continues its research into space-based defence. Furthermore, there is no evident common ground between the Americans and the Russians on strategic or intermediate-range nuclear weapons. The Soviet Union rejects the US view that its large ICBM force, with its high proportion of multiple independently-targeted re-entry vehicles (MIRVs) presents a 'first strike' threat to the United States. To the Americans a Soviet nuclear armoury of 1,398 ICBMs with between 6,000 and 7,000 warheads looks formidable in comparison with their own 1,030 missiles with 2,000 warheads, especially with a new Soviet missile (the SS-X-24) coming into operation in 1985, with up to ten MIRVs. It was for this reason that President Reagan was concerned to go ahead with the MX missile (also with up to ten MIRVs) before the Geneva talks were fully engaged.

Furthermore, the Soviet Union insists, not without some justification, that the British and French ICBMs (a total of 144 missiles with plans for MIRVs) should be taken into account.

The disagreement about intermediate range missiles will be as difficult to resolve. For the Americans and their allies 400 mobile SS-20s, each carrying three MIRVs and each with a reload, present a formidable threat to Western Europe. To the Russians, the 464 ground-launched cruise missiles and 108 Pershing II ballistic missiles which the US proposes to deploy in Western Europe (with the agreement of its NATO allies) pose a similar threat to the Soviet mainland. In addition the Soviet Union insists that all aspects of strategic arms control – space, intercontinental nuclear and intermediate-range nuclear – are so interconnected that no agreement in one is possible in isolation from the others; while the American view is that agreement in any one category of nuclear weapons should not be held up by failure to agree about others. There will clearly be no quick or easy solutions at Geneva.

During this early, near-term, phase it is important that the West – and specifically Western Europe – should begin to make some serious progress in the improvement of its non-nuclear defences to the point at which 'tactical' or 'battlefield' nuclear weapons become irrelevant. This can be most effectively achieved by the application of advanced technology to the battlefield. It is an area of military development in which the incidence of 'nay-sayers' is as great as in the debate about ballistic missile defence. The concept of deep-strike defence, employing the full range of new technological developments, is rejected for many of the same reasons which are deployed against the SDI: either it is not technologically feasible; or if it is, it is too expensive; or if it can be done and we can afford it, it is too provocative and 'destabilizing'. As with the doctrine of Mutual Assured Destruction – or, indeed, as part of it – there seems to be a determination to preserve strategic and tactical doctrines that have their roots firmly embedded in the familiar thought patterns of World War II.

If the research programme suggests that defence against nuclear missiles might be feasible, it will then be appropriate to move into a new, transitional phase. It can be assumed that the Soviet Union will have continued its own ballistic missile research and development programme; and somewhere around the turn of the century it might prove possible to move to a strategic relationship which will place greater reliance on defence and less on retaliation. At this stage it will be necessary for the United States to take account of the interests and preoccupations of its allies; and also of the implications for international agreements – notably the ABM Treaty with the Soviet Union. It will be important, too, to ensure that any defensive system that it is proposed to deploy suffers from none of the deficiencies now confidently foreshadowed by the opponents of SDI. The defensive systems themselves must be 'survivable', that is to say, not vulnerable to a first strike; otherwise the charge of 'destabilization' would be justified. Furthermore, they must not be so expensive that they could easily be outflanked by relatively inexpensive increments to the offensive capability of the other side.

During this period, close continuing contacts with the Soviet Union in the field of arms control would be of crucial importance. There would ideally be further verifiable reductions in offensive systems. The development and deployment of ballistic missile defences would proceed systematically and, to a large extent, bilaterally, that is to say in co-operation with the Soviet Union. Meanwhile the West would have to continue to enhance its conventional defensive capability. At this stage, too, it would also be necessary to begin to engage the other nuclear-weapons powers, as well as the 'near nuclear-weapons powers' in the process. This phase might well need to be spread over a period of twenty to twenty-five years.

It is at this stage that there is a significant difference of approach, even among those who support the Strategic Defence Initiative. President Reagan, in his original March 1983 speech, spoke of 'rendering these nuclear weapons impotent and obsolete' and of paving the way for 'arms control measures to eliminate the weapons themselves'. The official view of the US administration is that, in the long term, it will be possible to continue, as Paul Nitze has said, 'the reduction of nuclear weapons down to zero'. Others believe that it will always be necessary to retain a residual nuclear capability.

The official American position was set out by Nitze in an address to the National Academy of Sciences in April 1985, shortly after the resumption of the Geneva talks.

> With regard to preventing an arms race in space and terminating it on earth, contrary to Soviet assertions, that is exactly what our strategic concept envisages. The term 'arms race' connotes a runaway competition between two sides, with each piling weapon upon weapon in an unbridled manner. What we propose is just the opposite – a new transition to greater reliance on defensive systems, should new technologies prove feasible, managed jointly by the United States and the Soviet Union as a cooperative endeavor. Defenses would be introduced at a measured pace, in conjunction with progressively stricter limitations and reductions in offensive nuclear arms. The result would be that the two sides would have far fewer weapons which would use space as a medium for delivering nuclear destruction. The cooperative approach we foresee would be designed to maintain at all times control over the mix of offensive and defensive systems on both sides, and thereby increase the confidence of the sides in the effectiveness and stability of the deterrent balance.
>
> With regard to limiting and reducing nuclear arms, we have proposed deep reductions in strategic nuclear arms and the elimination of the entire class of US and Soviet longer-range INF missiles. Furthermore, we have expressed our commitment to the ultimate objective of eliminating all nuclear weapons. We are convinced that, if we are able in the future to deploy cost-effective defenses, they would provide incentives to movement toward that objective.

The arguments against the 'abolition' or 'elimination' of nuclear weapons are reasonably familiar. The principal objection is that they cannot be abolished. The knowledge of how to make nuclear weapons cannot be expunged from the collective human consciousness; and there is always the danger, in some international crisis, that one side or the other will decide that sooner than be dictated to, it will construct and use or threaten to use a nuclear weapon. Those who believe that nuclear weapons can, eventually, be eliminated claim, as Mr Paul Nitze did in his exposition of the new strategic concept, that, given the deployment of effective non-nuclear defences, cheating by the clandestine construction of nuclear delivery systems would have to be on such a large scale that it would be relatively easy to detect and counteract.

Even if this were true, it would still not deal with the contingency of the terrorist state, and such phenomena as the 'suitcase bomb'. However, even if it proved, in the long term, impossible to eliminate nuclear weapons entirely, there is general agreement among supporters of the Strategic Defence Initiative that nuclear weapons systems could be reduced to a level consistent with 'minimum deterrence', thus removing the instability posed by the potential threat of a first strike. Whether nuclear weapons continued to exist or not, the dangers of nuclear war would have been substantially reduced. This is, of course, a distant vision. It will not be realized, if it is realized at all, until well into the twenty-first century, and it will be formidably expensive. Yet, leaving aside the arguments, most of which lack substance, about the need to 'divert military spending into resources to combat starvation in the Third World', it is difficult to dissent from President Reagan's proposition that it is 'worth every investment to free the world from the threat of nuclear war'.

Those who oversimplify the profoundly complicated issue of the Strategic Defence Initiative with ill-considered objections to 'extending the arms race into space' seem to be prepared to ignore the fact that the confrontation between the superpowers already depends to a large extent upon the military exploration of space. None of the strategic nuclear weapons systems at present deployed could function effectively without the satellites now in orbit to provide early warning, reconnaissance, communications and target information. By the end of 1983, there were at least 2,000 military satellites in orbit, engaged in a wide range of military applications. Space already has a significant role in contemporary military strategy. It will inevitably continue to do so, since there is no existing international agreement regulating the military uses of space, with the exception of the 1967 Outer Space Treaty forbidding the placing in orbit of weapons of mass destruction. It would therefore be unwise for those involved with the development of military strategy to assume that space may not, before too long, prove to be the 'high ground' that is the traditional concern of the military planner. The Strategic Defence Initiative is too important and too complicated to be assailed with arguments derived from discredited strategic doctrines and based upon outmoded habits of strategic thought.

Appendix 1

President Reagan's Speech, 23 March 1983

Thank you for sharing your time with me tonight. The subject I want to discuss with you, peace and national security – is both timely and important – timely because I have reached a decision which offers a new hope for our children in the 21st century – a decision I will tell you about in a few minutes – and important because there is a very big decision that you must make for yourselves. This subject involves the most basic duty that any President and any people share – the duty to protect and strengthen the peace.

At the beginning of this year, I submitted to the Congress a defense budget which reflects my best judgment, and the best understanding of the experts and specialists who advise me, about what we and our allies must do to protect our people in the years ahead.

That budget is much more than a long list of numbers, for behind all the numbers lies America's ability to prevent the greatest of human tragedies and preserve our free way of life in a sometimes dangerous world. It is part of a careful, long-term plan to make America strong again after too many years of neglect and mistakes. Our efforts to rebuild America's defenses and strengthen the peace began two years ago, when we requested a major increase in the defense program. Since then, the amount of those increases we first proposed has been reduced by half through improvements in management and procurement and other savings. The budget request that is now before the Congress has been trimmed to the limits of safety. Further, deep cuts cannot be made without seriously endangering the security of the nation. The choice is up to the men and women you have elected to the Congress – and that means the choice is up to you.

Tonight I want to explain to you what this defense debate is all about, and why I am convinced that the budget now before the Congress is necessary, responsible, and deserving of your support. And I want to offer hope for the future.

But let me say first what the defense debate is not about. It is not about spending arithmetic. I know that in the last few weeks you've been bombarded with numbers and percentages. Some say we need only a 5 percent increase in defense spending. The so-called 'alternate budget', backed by liberals in the House of Representatives, would lower the figure to 2 to 3 percent, cutting our defense spending by 163 billion dollars over the next five years. The trouble with all these numbers is that they tell us little about the kind of defense program America needs, or the benefits in security and freedom that our defense effort buys for us.

What seems to have been lost in all this debate is the simple truth of how a defense budget is arrived at. It isn't done by deciding to spend a certain number of dollars. Those loud voices that are occasionally heard charging that the government is trying to solve a security problem by throwing money at it are nothing more than noise based on ignorance.

We start by considering what must be done to maintain peace, and review all the possible threats against our security. Then a strategy for strengthening peace and defending against those threats must be agreed upon. And finally our defense establishment must be evaluated to see what is necessary to protect against any or all the potential threats. The cost of achieving these ends is totaled up and the result is the budget for national defense.

There is no logical way you can say, let's spend *x* billion dollars less. You can only say, which part of our defense measures do we believe we can do without and still have security against all contingencies? Anyone in the Congress who advocates a percentage or specific dollar cut in defense spending should be made to say what part of our defenses he would eliminate, and he should be candid enough to acknowledge that his cuts mean cutting our commitments to allies or inviting greater risk or both.

The defense policy of the United States is based on a simple premise: the United States does not start fights. We will never be an aggressor. We maintain our strength in order to deter and defend against aggression – to preserve freedom and peace.

Since the dawn of the atomic age, we have sought to reduce the risk of war by maintaining a strong deterrent, and by seeking genuine arms control. 'Deterrence' means simply this: making sure any adversary who thinks about attacking the United States, or our allies, or our vital interests, concludes that the risks to him outweigh any potential gains. Once he understands that, he won't attack. We maintain the peace through our strength; weakness only invites aggression.

This strategy of deterrence has not changed. It still works. But what it takes to maintain deterrence has changed. It took one kind of military force to deter an attack when we had far more nuclear weapons than any other power; it takes another kind now that the Soviets, for example, have enough accurate and powerful nuclear weapons to destroy virtually all of our missiles on the ground. Now this is not to say the Soviet Union is planning to make war on us. Nor do I believe a war is inevitable – quite the contrary. But what must be recognized in that our security is based on being prepared to meet all threats.

There was a time when we depended on coastal forts and artillery batteries because, with the weaponry of that day, any attack would have had to come by sea. This is a different world and our defenses must be based on recognition and awareness of the weaponry possessed by other nations in the nuclear age.

We can't afford to believe we will never be threatened. There have been two World Wars in my lifetime. We didn't start them and, indeed, we did everything we could to avoid being drawn into them. But we were ill prepared for both –

had we been better prepared, peace might have been preserved.

For 20 years, the Soviet Union has been accumulating enormous military might. They didn't stop when their forces exceeded all requirements of a legitimate defensive capability. And they haven't stopped now.

During the past decade and a half, the Soviets have built up a massive arsenal of new strategic weapons – weapons that can strike directly at the United States.

As an example, the United States introduced its last new intercontinental ballistic missile, the Minuteman III, in 1969, and we are now dismantling our even older Titan missiles. But what has the Soviet Union done in these intervening years? Well, since 1969, the Soviet Union has built five new classes of ICBMs, and upgraded these eight times. As a result, their missiles are much more powerful and accurate than they were several years ago and they continue to develop more, while ours are increasingly obsolete.

The same thing has happened in other areas. Over the same period, the Soviet Union built four new classes of submarine-launched ballistic missiles and over sixty new missile submarines. We built two new types of submarine missiles and actually withdrew ten submarines from strategic missions. The Soviet Union built over two hundred new Backfire bombers, and their brand-new Blackjack bomber is now under development. We haven't built a new long-range bomber since our B-52s were deployed about a quarter of a century ago, and we've already retired several hundred of those because of old age. Indeed, despite what many people think, our strategic forces cost only about 15 percent of the defense budget.

Another example of what's happened: in 1978, the Soviets had 600 intermediate-range nuclear missiles based on land and were beginning to add the SS-20 – a new, highly accurate mobile missile, with three warheads. We had none. Since then the Soviets have strengthened their lead. By the end of 1979, when Soviet leader Brezhnev declared 'A balance now exists', the Soviets had over 800 warheads. We still had none.

Some freeze. At this time the Soviet Defense Minister Ustinov announced 'Approximate parity of forces continues to exist'. But the Soviets are still adding an average of three new warheads a week, and now have 1,300. These warheads can reach their targets in a matter of a few minutes. We still have none. So far, it seems that the Soviet definition of parity is a box score of 1,300 to nothing, in their favor.

So, together with our NATO allies, we decided in 1979 to deploy new weapons, beginning this year, as a deterrent to their SS-20s and as an incentive to the Soviet Union to meet us in serious arms control negotiations. We will begin that deployment late this year. At the same time, however, we are willing to cancel our program if the Soviets will dismantle theirs. This is what we call a zero-zero plan. The Soviets are now at the negotiating table – and I think it's fair to say that without our planned deployments, they wouldn't be there.

Now let's consider conventional forces. Since 1974, the United States has produced 3,050 tactical combat aircraft. By contrast, the Soviet Union has produced twice as many. When we look at attack submarines, the United States has produced 27, while the Soviet Union has produced 61. For armored vehicles including tanks, we have produced 11,200. The Soviet Union has produced 54,000, a nearly 5 to 1 ratio in their favor. Finally, with artillery, we have produced 950 artillery and rocket launchers while the Soviets have produced more than 13,000, a staggering 14 to 1 ratio.

There was a time when we were able to offset superior Soviet numbers with higher quality. But today they are building weapons as sophisticated and modern as our own.

As the Soviets have increased their military power, they have been emboldened to extend that power. They are spreading their military influence in ways that can directly challenge our vital interests and those of our allies. The following aerial photographs, most of them secret until now, illustrate this point in a crucial area very close to home – Central America and the Caribbean basin. They are not dramatic photographs but I think they help give you a better understanding of what I'm talking about.

This Soviet intelligence collection facility less than 100 miles from our coast is the largest of its kind in the world. The acres and acres of antennae fields and intelligence monitors are targeted on key US military installations and sensitive activities. The installation, in Lourdes, Cuba, is manned by 1,500 Soviet technicians, and the satellite ground station allows instant communications with Moscow. This 28-square-mile facility has grown by more than 60 percent in size and capability during the past decade.

In western Cuba, we see this military airfield and its complement of modern Soviet-built MIG-23 aircraft. The Soviet Union uses this Cuban airfield for its own long-range reconnaissance missions, and earlier this month two modern Soviet anti-submarine warfare aircraft began operating from it. During the past two years, the levels of Soviet arms exports to Cuba can only be compared to the levels reached during the Cuban missile crisis 20 years ago.

This third photo, which is the only one in this series that has previously made public, shows Soviet military hardware that has made its way to Central America. This airfield with its MI-8 helicopters, anti-aircraft guns, and protected fighter sites is one of a number of military facilities in Nicaragua which has received Soviet equipment funneled through Cuba, and reflects the massive military build-up going on in that country.

On the small island of Grenada, at the southern end of the Caribbean chain, the Cubans, with Soviet financing and backing, are in the process of building an airfield with a 10,000-foot runway. Grenada doesn't even have an air force. Who is it intended for? The Caribbean is a very important passageway for our international commerce and military lines of communication. More than half of all American oil imports now pass through the Caribbean. The rapid build-up of Grenada's military potential is unrelated to any conceivable threat to this island country of under 110,000 people, and totally at odds with the pattern of other eastern Caribbean states, most of which are unarmed. The Soviet–Cuban militarization of Grenada, in short, can only be seen as power projection into the region, and it is in this important economic and strategic area that we are trying to help the governments of El Salvador, Costa Rica, Honduras, and others in their struggles for democracy against guerillas supported through Cuba and Nicaragua.

These pictures only tell a small part of the story. I wish I could show you more without compromising our most sensitive intelligence sources and methods. But the Soviet Union is also supporting Cuban military forces in Angola and Ethiopia. They have bases in Ethiopia and South Yemen near the Persian Gulf oil fields. They have taken over the port we built at Cam Ranh Bay in Vietnam, and now, for the first time in history, the Soviet Navy is a force to be reckoned with in the South Pacific.

Some people may still ask: would the Soviets ever use their formidable military power? Well, again, can we afford to believe they won't? There is Afghanistan; and in Poland, the Soviets denied the will of the people and, in so doing, demonstrated to the world how their military power could also be used to intimidate.

The final fact is that the Soviet Union is acquiring what can only be considered an offensive military force. They have continued to build far more intercontinental ballistic missiles than they could possibly need simply to deter an attack. Their conventional forces are trained and equipped not so much to defend against an attack as they are to permit sudden, surprise offensives of their own.

Our NATO allies have assumed a great defense burden, including the military draft in most countries. We are working with them and our other friends around the world to do more. Our defensive strategy means we need military forces that can move very quickly – forces that are trained and ready to respond to any emergency.

Every item in our defense program – our ships, our tanks, our planes, our funds for training and spare parts – is intended for one all-important purpose – to keep the peace. Unfortunately, a decade of neglecting our military forces had called into question our ability to do that.

When I took office in January 1981, I was appalled by what I found: American planes that could not fly and American ships that could not sail for lack of spare parts and trained personnel and insufficient fuel and ammunition for essential training. The inevitable result of all this was poor morale in our armed forces, difficulty in recruiting the brightest young Americans to wear the uniform, and difficulty in convincing our most experienced military personnel to stay on.

There was a real question, then, about how well we could meet a crisis. And it was obvious that we had to begin a major modernization program to ensure we could deter aggression and preserve the peace in the years ahead.

We had to move immediately to improve the basic readiness and staying power of our conventional forces, so they could meet – and therefore help to deter – a crisis. And it was obvious that we had to make up for lost years of investment by moving forward with a long-term plan to prepare our forces to counter the military capabilities our adversaries were developing for the future.

I know that all of you want peace and so do I. I know too that many of you seriously believe that a nuclear freeze would further the cause of peace. But a freeze now would make us less, not more, secure and would raise, not reduce, the risks of war. It would be largely unverifiable and would seriously undercut our negotiations on arms reduction. It would reward the Soviets for their massive military buildup while preventing us from modernizing our ageing and increasingly vulnerable forces. With their present margin of superiority, why should they agree to arms reductions knowing that we were prohibited from catching up?

Believe me, it wasn't pleasant for someone who had come to Washington determined to reduce government spending, but we had to move forward with the task of repairing our defenses or we would lose our ability to deter conflict now and in the future. We had to demonstrate to any adversary that aggression could not succeed, and that the only real solution was substantial, equitable, and effectively verifiable arms reduction – the kind we're working for right now in Geneva.

Thanks to your strong support, and bipartisan support from the Congress, we began to turn things around. Already, we are seeing some very encouraging results. Quality recruitment and retention are up, dramatically – more high school graduates are choosing military careers and more experienced career personnel are choosing to stay. Our men and women in uniform at last are getting the tools and training they need to do their jobs.

Ask around today, especially among our young people, and I think you'll find a whole new attitude towards serving their country. This reflects more than just better pay, equipment, and leadership. You, the American people, have sent a signal to these young people that it is once again an honor to wear the uniform. That's not something you measure in a budget, but it is a very real part of our nation's strength.

It will take us longer to build the kind of equipment we need to keep peace in the future, but we've made a good start.

We have not built a new long-range bomber for 21 years. Now we're building the B-1. We had not launched one new strategic submarine for 17 years. Now, we're building one Trident submarine a year. Our land-based missiles are increasingly threatened by the many huge, new Soviet ICBMs. We are determining how to solve that problem. At the same time, we are working in the START and INF negotiations, with the goal of achieving deep reductions in the strategic and intermediate nuclear arsenals of both sides.

We have also begun the long-needed modernization of our conventional forces. The Army is getting its first new tank in 20 years. The Air Force is modernizing. We are rebuilding our Navy which shrank from about 1,000 in the late 1960s to 453 ships during the 1970s. Our nation needs a superior Navy to support our military forces and vital interests overseas. We are now on the road to achieving a 600-ship Navy and increasing the amphibious capabilities of our Marines who are now serving the cause of peace in Lebanon. And we are building a real capability to assist our friends in the vitally important Indian Ocean and Persian Gulf region.

This adds up to a major effort, and it is not cheap. It comes at a time when there are many other pressures on our budget, and when the American people have already had to make major sacrifices during the recession. But we must not be misled by those who would make defense once again the scapegoat of the Federal budget.

The fact is that in the past few decades we have seen a major shift in how we spend the taxpayer's dollar. Back in 1955, payments to individuals took up only 20 percent of the Federal budget. For nearly three decades, these payments steadily increased and this year will account for 49 percent of the budget. By contrast, in 1955, defense took up more than half of the Federal budget. By 1980, this spending had fallen to a low of 23 percent. Even with the increase I am requesting this year, defense will still amount to only 28 percent of the budget.

The calls for cutting back the defense budget come in nice simple arithmetic. They're the same kind of talk that led the democracies to neglect their defenses in the 1930s and invited the tragedy of World War II. We must not let that grim chapter of history repeat itself through apathy or neglect.

This is why I am speaking to you tonight – to urge you to tell your Senators and Congressman that you know we must continue to restore our military strength.

If we stop in midstream, we will send a signal of decline, of lessened will, to friends and adversaries alike.

Free people must voluntarily, through open debate and democratic means, meet the challenge that totalitarians pose by compulsion.

It is up to us, in our time, to choose, and choose wisely, between the hard but necessary task of preserving peace and freedom and the temptation to ignore our duty and blindly hope for the best while the enemies of freedom grow stronger day by day.

The solution is well within our grasp. But to reach it, there is simply no alternative but to continue this year, in this budget, to provide the resources we need to preserve the peace and guarantee our freedom.

Thus far tonight I have shared with you my thoughts on the problems of national security we must face together. My predecessors in the Oval Office have appeared before you on other occasions to describe the threat posed by Soviet power and have proposed steps to address that threat. But since the advent of nuclear weapons, those steps have been increasingly directed toward deterrence of aggression through the promise of retaliation. This approach to stability through offensive threat has worked. We and our allies have succeeded in preventing nuclear war for three decades. In recent months, however, my advisors, including in particular the Joint Chiefs of Staff, have underscored the necessity to break out of a future that relies solely upon offensive retaliation for our security.

Over the course of these discussions, I have become more and more deeply convinced that the human spirit must be capable of rising above dealing with other nations and human beings by threatening their existence. Feeling this way, I believe we must thoroughly examine every opportunity for reducing tensions and for introducing greater stability into the strategic calculus on both sides. One of the most important contributions we can make is, of course, to lower the level of all arms, and particularly nuclear arms. We are engaged right now in several negotiations with the Soviet Union to bring about a mutual reduction of weapons. I will report to you a week from tomorrow my thoughts on that score. But let me just say I am totally committed to this course.

If the Soviet Union will join with us in our effort to achieve major arms reduction we will have succeeded in stabilizing the nuclear balance. Nevertheless it will still be necessary to rely on the specter of retaliation – on mutual threat, and that is a sad commentary on the human condition.

Would it not be better to save lives than to avenge them? Are we not capable of demonstrating our peaceful intentions by applying all our abilities and our ingenuity to achieving a truly lasting stability? I think we are – indeed, we must!

After careful consultation with my advisors, including the

Joint Chiefs of Staff, I believe there is a way. Let me share with you a vision of the future which offers hope. It is that we embark on a mission to counter the awesome Soviet missile threat with measures that are defensive. Let us turn to the very strengths in technology that spawned our great industrial base and that have given us the quality of life we enjoy today.

What if free people could live secure in the knowledge that their security did not rest on the threat of instant US retaliation to deter a Soviet attack; that we could intercept and destroy strategic ballistic missiles before they reached our own soil or that of our allies?

I know this is a formidable technical task, one that may not be accomplished before the end of this century. Yet, current technology has attained a level of sophistication where it is reasonable for us to begin this effort. It will take years, probably decades, of effort on many fronts. There will be failures and setbacks just as there will be successes and breakthroughs. And as we proceed we must remain constant in preserving the nuclear deterrent and maintaining a solid capability for flexible response. But is it not worth every investment necessary to free the world from the threat of nuclear war? We know it is.

In the meantime, we will continue to pursue real reductions in nuclear arms, negotiating from a position of strength that can be ensured only be modernizing our strategic forces. At the same time, we must take steps to reduce the threat of a conventional military conflict escalating to nuclear war by improving our non-nuclear capabilities. America does possess – now – the technologies to attain very significant improvements in the effectiveness of our conventional, non-nuclear forces. Proceeding boldly with these new technologies, we can significantly reduce any incentive that the Soviet Union may have to threaten attack against the United States or its allies.

As we pursue our goal of defensive technologies, we recognize that our allies rely upon our strategic offensive powers to deter attacks against them. Their vital interests and ours are inextricably linked – their safety and ours are one.

And no change in technology can or will alter that reality. We must and shall continue to honor our commitments.

I clearly recognize that defensive systems have limitations and raise certain problems and ambiguities. If paired with offensive systems, they can be viewed as fostering an aggressive policy, and no one wants that.

But with these considerations firmly in mind, I call upon the scientific community who gave us nuclear weapons to turn their great talents to the cause of mankind and world peace; to give us the means of rendering these nuclear weapons impotent and obsolete.

Tonight, consistent with our obligations under the ABM Treaty and recognizing the need for close consultation with our allies, I am taking an important first step. I am directing a comprehensive and intensive effort to define a long-term research and development program to begin to achieve our ultimate goal of eliminating the threat posed by strategic nuclear missiles. This would pave the way for arms control measures to eliminate the weapons themselves. We seek neither military superiority nor political advantage. Our only purpose – one all people share – is to search for ways to reduce the danger of nuclear war.

My fellow Americans, tonight we are launching an effort which holds the promise of changing the course of human history. There will be risks, and results take time. But with your support, I believe we can do it.

Appendix 2
The ABM Treaty

During a visit to Moscow by President Nixon on 22–29 May 1972, a treaty on the limitation of anti-ballistic missile systems, an interim agreement of certain measures with respect to the limitation of strategic offensive arms and six different co-operation agreements were signed by the two sides. The *Treaty on the Limitation of Anti-Ballistic Missile Systems*, signed on 26 May 1972, was the culmination of over two years of talks on strategic arms limitation (SALT) held between the two governments. The treaty, also known as SALT I, and concluded for an unlimited duration, consisted of sixteen articles, including the following:

Art. 1. (1) Each party undertakes to limit anti-ballistic missile (ABM) systems and to adopt other measures in accordance with the provisions of this treaty.

(2) Each party undertakes not to deploy ABM systems for a defence of the territory of its country and not to provide a base for such a defence, and not to deploy ABM systems for defence of an individual region except as provided for in Article 3 of this treaty.

Art. 2. (1) For the purpose of this treaty an ABM system is a system to counter strategic ballistic missiles or their elements in flight trajectory currently consisting of:

(*a*) ABM interceptor missiles, which are interceptor missiles constructed and deployed for an ABM role, or of a type tested in an ABM mode;

(*b*) ABM launchers, which are launchers constructed and deployed for launching ABM interceptor missiles; and

(*c*) ABM radars, which are radars constructed and deployed for an ABM role, or of a type tested in an ABM mode.

(2) The ABM system components listed in paragraph 1 of this article include those which are: (*a*) operational; (*b*) under construction; (*c*) undergoing testing; (*d*) undergoing overhaul, repair or conversion; or (*e*) mothballed.

Art. 3. Each party undertakes not to deploy ABM systems or their components except that:

(*a*) Within one ABM system deployment area having a radius of 150 kilometres and centred on the party's national capital, a party may deploy: (1) No more than 100 ABM launchers and no more than 100 ABM interceptor missiles at launch sites; and (2) ABM radars within no more than six ABM radar complexes, the area of each complex being circular and having a diameter of no more than three kilometres; and

(*b*) within one ABM system deployment area having a radius of 150 kilometres and containing ICBM silo launchers, a party may deploy: (1) No more than 100 ABM interceptor missiles at launch sites; (2) two large phased array ABM radars comparable in potential to corresponding ABM radars operational or under construction on the date of signature of the treaty in an ABM system deployment area containing ICBM silo launchers; and (3) no more than 18 ABM radars each having a potential less than the potential of the smaller of the above-mentioned two large phased-array ABM radars.

Art. 4. The limitations provided for in Article 3 shall not apply to ABM systems or their components for development or testing, and located within current or additionally agreed test ranges. Each party may have no more than a total of 15 ABM launchers at test ranges.

Art. 5. (1) Each party undertakes not to develop, test, or deploy ABM systems or components which are sea-based, space-based or mobile land-based.

(2) Each party undertakes not to develop, test or deploy ABM launchers for launching more than one ABM interceptor missile at a time from each launcher, nor to modify deployed launchers to provide them with such a capability, nor to develop, test, or deploy automatic or semi-automatic or other similar systems for rapid reload of ABM launchers.

Art. 6. To enhance assurance of the effectiveness of the limitations on ABM systems and their components provided by this treaty, each party undertakes:

(*a*) Not to give missiles, launchers, or radars, other than ABM interceptor missiles, ABM launchers, or ABM radars, capabilities to counter strategic ballistic missiles or their elements in flight trajectory, and not to test them in an ABM mode; and

(*b*) not to deploy in the future radars for early warning of strategic ballistic missile attack except at locations along the periphery of its national territory and oriented outwards.

Art. 8. ABM systems or their components in excess of the numbers or outside the areas specified in this treaty, as well as ABM systems or their components prohibited by this treaty, shall be destroyed or dismantled under agreed procedures within the shortest possible agreed period of time.

Art. 9. To assure the viability and effectiveness of this treaty, each party undertakes not to transfer to other states, and not to deploy outside its national territory, ABM systems or their components limited by this treaty.

Art. 12. (1) For the purpose of providing assurance of compliance with the provisions of this treaty, each party shall use national technical means of verification at its disposal in a manner consistent with generally recognized principles of international law.

(2) Each party undertakes not to interfere with the national technical means of verification of the other party operating in accordance with Paragraph (1) of this article.

(3) Each party undertakes not to use deliberate concealment measures which impede verification by national technical means of compliance with the provisions of this treaty. This obligation shall not require changes in current construction, assembly, conversion or overhaul practices.

After approval by the US Senate, President Nixon ratified the treaty on 29–30 September 1972; the Supreme Soviet of the USSR approved it and President Podgorny ratified it on 29 September; and instruments of ratification were exchanged in Washington on 3 October 1972.

Appendix 3
The Outer Space Treaty

It was announced at the United Nations on 8 December 1966, that agreement had been reached on the first international treaty governing space exploration. The treaty, officially designated the *Treaty on the Principles of the Activity of States in the Exploration and Use of Outer Space Including the Moon and Other Celestial Bodies*, was unanimously approved by the UN General Assembly on 19 December 1966. It was signed by the United States, the Soviet Union, and the United Kingdom in their respective capitals on 27 January 1967, and by the date of its entry into force (10 October 1967), it had been signed in Washington, London or Moscow by a total of ninety-three nations.

The operative articles of the treaty read as follows:

Art. 1. The exploration and use of outer space, including the moon and other celestial bodies, shall be carried out for the benefit and in the interests of all countries, irrespective of their degree of economic or scientific development, and shall be the province of all mankind.

Outer space, including the moon and other celestial bodies, shall be free for exploration and use by all states without discrimination of any kind, on a basis of equality and in accordance with international law, and there shall be free access to all areas of celestial bodies.

There shall be freedom of scientific investigation in outer space, including the moon and other celestial bodies, and states shall facilitate and encourage international co-operation in such investigation.

Art. 2. Outer space, including the moon and other celestial bodies, is not subject to national appropriation by claim of sovereignty, by means of use or occupation or by any other means.

Art. 3. States parties to the treaty shall carry on activities in the exploration and use of outer space, including the moon and other celestial bodies, in accordance with international law, including the Charter of the United Nations, in the interest of maintaining international peace and security and promoting international co-operation and understanding.

Art. 4. States parties to the treaty undertake not to place in orbit around the earth any objects carrying nuclear weapons or any other kinds of weapons of mass destruction, install such weapons on celestial bodies, or station such weapons in outer space in any other manner.

The moon and other celestial bodies shall be used by all states parties to the treaty exclusively for peaceful purposes. The establishment of military bases, installations and fortifications, the testing of any type of weapons, and the conduct of military manoeuvres on celestial bodies shall be forbidden. The use of military personnel for scientific research or for any other peaceful purposes shall not be prohibited. The use of any equipment or facility necessary for peaceful exploration of the moon and other celestial bodies shall also not be prohibited.

Art. 5. States parties to the treaty shall regard astronauts as envoys of mankind in outer space and shall render to them all possible assistance in the event of accident, distress or emergency landing on the territory of another state party or on the high seas. When astronauts make such a landing, they shall be safely and promptly returned to the state of registry of their space vehicle.

In carrying on activities in outer space and on celestial bodies, the astronauts of one party shall render all possible assistance to the astronauts of other states parties.

States parties to the treaty shall immediately inform the other states parties to the treaty or the Secretary-General of the United Nations of any phenomena they discover in outer space, including the moon and other celestial bodies, which could constitute a danger to the life or health of astronauts.

Art. 6. States parties to the treaty shall bear international responsibility for national activities in outer space, including the moon and other celestial bodies, whether such activities are carried on by governmental agencies or by non-governmental entities, and for assuring that national activities are carried out in conformity with the provisions set forth in the present treaty.

The activities of non-governmental entities in outer space, including the moon and other celestial bodies, shall require authorization and continuing supervision by the state concerned. When activities are carried on in outer space, including the moon and other celestial bodies, by an international organization, responsibility for compliance with this treaty shall be borne both by the international organization and by the states parties to the treaty participating in such organization.

Art. 7. Each state party to the treaty that launches or procures the launching of an object into outer space, including the moon and other celestial bodies, and each state party from whose territory or facility an object is launched, is internationally liable for damage to another state party to the treaty or to its natural or juridical persons by such object or its component parts on the earth, in airspace, or in outer space, including the moon and other celestial bodies.

Art. 8. A state party to the treaty on whose registry an object launched into outer space is carried shall retain jurisdiction and control over such object, and over any personnel thereof, while in outer space or on a celestial body. Ownership of objects launched into outer space, including objects landed or constructed on a celestial body, and of their component parts, is not affected by their presence in outer space, including the body, or by their return to the earth. Such objects or component parts found beyond the limits of the state party to the treaty on whose registry they are carried shall be returned to that state, which shall, upon request, furnish identifying data prior to their return.

Art. 9. In the exploration and use of outer space, including the moon and other celestial bodies, states parties to the treaty shall be guided by the principle of co-operation and mutual assistance and shall conduct all their activities in outer space, including the moon and other celestial bodies, with due regard to the corresponding interests of all other states parties to the treaty.

States parties to the treaty shall pursue studies of outer space, including the moon and other celestial bodies, and conduct exploration of them so as to avoid their harmful contamination and also adverse changes in the environment of the earth resulting from the introduction of extra-terrestrial matter, and, where necessary, shall adopt appropriate measures for this purpose.

If a state party to the treaty has reason to believe that an activity or experiment planned by it or its nationals in outer space, including the moon and other celestial bodies, would cause potentially harmful interference with activities of other state parties in the peaceful exploration and use of outer space, including the moon and other celestial bodies, it shall undertake appropriate international consultations before proceeding with any such activity or experiment.

A state party to the treaty which has reason to believe that an activity or experiment planned by another state party in outer space, including the moon and other celestial bodies, would cause potentially harmful interference with activities in the peaceful exploration and use of outer space, including the moon and other celestial bodies, may request consultation concerning the activity or experiment.

Art. 10. In order to promote international co-operation in the exploration and use of outer space, including the moon and other celestial bodies, in conformity with the purposes of this treaty, the states parties to the treaty shall consider on a basis of equality any requests by other states parties to the treaty to be afforded an opportunity to observe the flight of space objects launched by those states. The nature of such an opportunity for observation and the conditions under which it could be afforded shall be determined by agreement between the states concerned.

Art. 11. In order to promote international co-operation in the peaceful exploration and use of outer space, states parties to the treaty conducting activities in outer space, including the moon and other celestial bodies, agree to inform the Secretary-General of the United Nations as well as the public and the international scientific community, to the greatest extent feasible and practicable, of the nature, conduct, locations and results of such activities. On receiving the said information, the UN Secretary-General should be prepared to disseminate it immediately and effectively.

Art. 12. All stations, installations, equipment and space vehicles on the moon and other celestial bodies shall be open to representatives of other states parties to the treaty, on a basis of reciprocity. Such representatives shall give reasonable advance notice of a projected visit, in order that appropriate consultations may be held and that maximum precautions may be taken to assure safety and to avoid interference with normal operations in the facility to be visited.

Art. 13. The provisions of this treaty shall apply to the activities of states parties to the treaty in the exploration and use of outer space, including the moon and other celestial bodies, whether such activities are carried on by a single state party to the treaty or jointly with other states, including cases where they are carried on within the framework of international inter-governmental organizations.

Any practical questions arising in connexion with activities carried on by international inter-governmental organizations in the exploration and use of outer space, including the moon and other celestial bodies, shall be resolved by the states parties to the treaty either with the appropriate international organization or with one or more state members of that international organization, which are parties to this treaty.

Art. 14. (1) This treaty shall be open to all states for signature. Any state which does not sign this treaty before its entry into force in accordance with Paragraph 3 of this article may accede to it at any time.

(2) This treaty shall be subject to ratification by signatory states. Instruments of ratification and of accession shall be deposited with the Governments of the Union of Soviet Socialist Republics, the United Kingdom of Great Britain and Northern Ireland and the United States of America, which are hereby designated the depositary governments.

(3) This treaty shall enter into force upon the deposit of instruments of ratification by five governments, including the governments designated as depositary governments.

(4) For states whose instruments of ratification or accession are deposited subsequent to the entry into force of this treaty, it shall enter into force on the date of the deposit of their instruments of ratification or accession.

(5) The depositary governments shall promptly inform all signatory and acceding states of the date of each signature, the date of deposit of each instrument of ratification of and accession to this treaty, the date of its entry into force, and other notices.

(6) This treaty shall be registered by the depositary governments pursuant to Article 102 of the UN Charter.

Art. 15. Any state party to the treaty may propose amendments to this treaty. Amendments shall enter into force for each state party to the treaty accepting the amendments upon their acceptance by a majority of the states parties to the treaty, and thereafter for each remaining state party to the treaty on the date of acceptance by it.

Art. 16. Any state party to the treaty may give notice of its withdrawal from the treaty one year after its entry into force by written notification to the depositary governments. Such withdrawal shall take effect one year from the date of receipt of this notification.

Notes

[1] Extracts from 'Peace and National Security', President Reagan's Televised Address to the Nation, 23 March 1983; full text in *Realism, Strength, Negotiation: Key Foreign Policy Statements of the Reagan Administration*, US Department of State, Bureau of Public Affairs, 1984.

[2] *Pravda*, 27 March 1983.

[3] *A Space-Based Anti-Missile System with Directed-Energy Weapons: Strategic, Legal and Political Implications*, Institute of Space Research, USSR Academy of Sciences, 1984. For additional extracts, see *Survival*, vol. XXVII, no. 2, March/April 1985.

[4] *The Times*, 6 May 1985.

[5] For the views of the Union of Concerned Scientists, see John Tirman (ed.), *A Space-Based Missile Defense: The Fallacy of Star Wars*, Vintage, 1984.

[6] 'European Defence and Political Co-operation', address to the Royal Institute of International Affairs, *The Times*, 14 March 1985.

[7] TASS, 10 February 1985.

[8] For an analysis of the doctrine of massive retaliation and its relevance to American nuclear policy, see Urs Schwartz, *American Strategy: A New Perspective*, Heinemann, London, 1967; and Gregg Herken, *The Winning Weapon: The Atomic Bomb and the Cold War*, Alfred A. Knopf, New York, 1980.

[9] See, for example, Henry A. Kissinger, 'The Future of NATO', in Kenneth A. Myers (ed.), *NATO: The Next Thirty Years*, Croom Helm, London, 1980.

[10] For the full text of NATO's 'two-track' decision, to install new missiles whilst pursuing arms control negotiations with the Soviet Union, see *NATO Final Communiques, Vol. II, 1975–1980*, NATO Information Service, 1110 Brussels, 1981 ('The Ministers have decided to pursue these two parallel and complementary approaches, in order to avert an arms race in Europe caused by the Soviet TNF build-up, yet preserve the viability of NATO's strategy of deterrence and defence and thus maintain the security of its member states').

[11] For an assessment of this strategic equation, see *Can America Catch Up? The US–Soviet Military Balance*, Committee on the Present Danger, Washington DC, 1984; and Norman Polmar, *Strategic Weapons: An Introduction*, Crane Russak, New York, 1982.

[12] See, for example, R.J. Rummel, 'Will the Soviet Union soon have a First-Strike Capability?', in *Orbis*, Fall 1976; Richard Pipes, 'Why the Soviet Union Thinks it could Fight and Win a Nuclear War', in *Commentary*, July 1977; Richard Burt, 'The Scope and Limits of SALT', in *Foreign Affairs*, July 1978.

[13] The *Sunday Times*, 23 December 1984. See also Gerald Frost, 'Mrs Thatcher's "Star Wars" Fantasy', in the *Wall Street Journal*, 27 February 1985.

[14] Paul Nitze, 'The Objectives of Arms Control', the 1985 Alastair Buchan Memorial Lecture at the International Institute for Strategic Studies, London, 28 March 1985.

[15] See, for example, Daniel O. Graham, *We Must Defend America*, Regnery Gateway, Chicago, 1983; and the findings of Lt. Gen. Graham's project, *High Frontier: A New National Strategy*, The Heritage Foundation, Washington DC, 1982. See also Zbigniew Brzezinski, Robert Jastrow and Max M. Kampelman, 'Search for Security: the Case for the Strategic Defense Initiative', in the *International Herald Tribune*, 28 January 1985.

[16] *Ballistic Missile Defenses and US National Security*, the Future Security Strategy Society, October 1983.

[17] James C. Fletcher, 'Technologies for Strategic Defensc', in *Issues in Science and Technology*, Fall 1984.

[18] For a detailed examination of these points, see W. Bruce Weinrod (ed.), *Assessing Strategic Defense: Six Roundtable Discussions*, The Heritage Foundation, Washington DC, 1985.

[19] *A Space-Based Missile Defense*, supra. Robert Jastrow, 'Reagan versus the Scientists: Why the President is Right about Missile Defence', in *Commentary*, vol. 77, no. 1, January 1984; and 'The War Against "Star Wars" ', in *Commentary*, vol. 78, no. 6, December 1984.

[20] For full figures, see *The Military Balance, 1984–1985*, International Institute for Strategic Studies, London, 1984. See also David B. Rivkin and Manfred R. Hamm, 'In Strategic Defense Moscow is far ahead', Backgrounder no. 409, The Heritage Foundation, Washington DC, February 1985; and Daniel Goure and Gordon McCormick, 'Soviet Strategic Defense: The Neglected Dimension of the US–Soviet Balance', in *Orbis*, Spring 1980.

[21] Thc *New York Times*, 17 January 1985.

[22] The 'Appeal to all Scientists of the World', signed by 244 Soviet scientists, was published by TASS on 9 April 1983.

[23] An extensive exposition is to be found in Bhupendra Jasani (ed.), *Space Weapons: The Arms Control Dilemma*, Stockholm International Peace Research Institute, Taylor & Francis, London, 1984.

[24] Statement before the Subcommittee on Research and Development of the Committee on Armed Services, 1 March 1984.

[25] *Star Wars Strategy: The Implications for the Europeans of the Atlantic Alliance*, vol. I, European Institute for Peace and Security, June 1984. See also Werner Kaltefleiter, *The Strategic Defence Initiative: Some Implications for Europe*, Institute for European Defence and Strategic Studies, 1985; and Hubertus G. Hoffmann, 'A Missile Defence for Europe', in *Strategic Review*, Summer 1984.

[26] For an analysis of the potential role of new technologies in strategic and conventional defence, see the *Sunday Times*, 21 October 1984; David Greenwood, 'Strengthening Conventional Deterrence Doctrine: New Technologies and Resources', in *NATO Review*, vol. 32, no. 4, August 1984; General Bernard Rogers, 'Follow-On Forces Attack (FOFA): Myths and Realities', in *NATO Review*, vol. 32, no. 6, December 1984; and R. V. Jones, *Future Conflict and New Technology*, Washington Paper no. 88, Praeger Publishers, 1981.

[27] 'Defence and Security in the Nuclear Age', address to the Royal United Services Institute, 15 March 1985: 'Impressions can be created by words as well as deeds. Policies, aims, visions – all these can and must be clearly stated. Without the approval of an informed public, the governments of the West are wasting their breath . . . . Words and dreams cannot by themselves justify what the Prime Minister described to the United Nations as the "perilous pretence" that a better system than nuclear deterrence is within reach at the present time.'

[28] The link between strategic defence and stronger conventional forces in President Reagan's speech of March 1983 has been repeatedly underlined by Dr George Keyworth, the Science Advisor to the President, for example in his address to the US Air Force National Convention in Washington DC, 17 September 1984.

[29] General Bernard Rogers, 'Sword and Shield: ACE Attack of Warsaw Pact Follow-on Forces', *NATO's 16 Nations*, vol. 28, February–March 1983, p. 16

[30] General Bernard Rogers, 'Greater Flexibility for NATO's Flexible Response', *Strategic Review*, Spring 1983.

[31] Dr Manfred Wörner, press conference (Bonn, 21 May 1982) on release of an Expert Group Study on a NATO 'Conventional Force Modernization Program'.

[32] NATO Military Committee Document 14/2.
[33] Report by Senator Sam Nunn (D-GA) to the Committee on Armed Services, United States Senate, 13 May 1982.
[34] Soviet Army Echelonment and Objectives prepared for Centag Commanders Conference by the US Defense Nuclear Agency, October 1978.
[35] General Donn A. Starry, 'Extending the Battlefield', *Military Review*, March 1981.
[36] Charles W. Corddry, *Baltimore Sun*.
[37] *The Economist*, 31 July 1982.

Index

Note: italicized page numbers denote diagrams on page cited.